OUTLINE OF CAT ANATOMY
with reference to the human

Illustrations and text by Stephen G. Gilbert

with the collaboration of Cheralea Gilbert

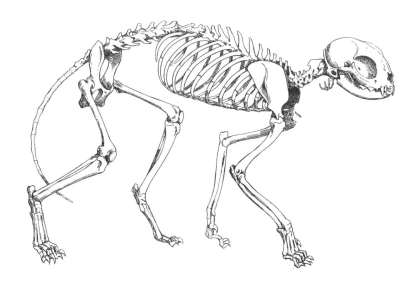

University of Washington Press   Seattle and London

Library of Congress Cataloging-in-Publication Data
Gilbert, Stephen G.
    Outline of cat anatomy, with reference to the human / illustrations
and text by Stephen G.
    Gilbert; with the collaboration of Cheralea Gilbert.
        p. cm.
    Includes bibliographical references.
    ISBN 0-295-97818-X
    1. Cats—Anatomy—Atlases. 2. Cats—Dissection—Atlases.
    I. Gilbert, Cheralea. II. Title.
    QL813.C38G55   1999
    571.3'1752—dc21   99-14036
                                                       CIP

The paper used in this publication meets the minimum requirements of
American National Standard for Information Sciences—Permanence of
Paper for Printed Library Materials, ANSI Z39.48-1984. ♾

CONTENTS

Preface   v

The Skeleton   3

The Muscles   14

The Respiratory and Digestive Systems   28

The Urogenital System   38

The Circulatory System   48

The Nervous System   62

The Eye   80

The Ear   84

Definitions of Descriptive Terms   86

Bibliography   87

Index   88

PREFACE

This book is an abridged and modified version of my *Pictorial Anatomy of the Cat,* now in its twelfth printing, and it is designed as a dissection guide to supplement textbooks used in introductory courses in human and mammalian anatomy. Such courses are usually offered as an introduction to the study of human anatomy, but in many cases students do not have access to human cadavers and the cat is used instead. I have therefore included illustrations of human anatomy, which may be compared with the dissection of the cat, and have for the most part employed anatomical terms used in human anatomy.

The use of the cat as a subject for dissection and comparison with the human has a long and interesting history. The first publication devoted to the comparative anatomy of the cat and the human was Hercule Straus-Durckheim's *Anatomie descriptive et comparative du chat,* published in Paris in 1845. Straus-Durckheim's work, which is limited to an account of the skeleton and the muscles, includes an atlas of large copperplate engravings and two large volumes, which together contain more than a thousand pages of text. His illustrations are a tour de force of meticulous craftsmanship; his exhaustively detailed text compares every bone, ligament, and muscle of the cat with its homologue in humans and other mammals. "I think I have omitted nothing," Straus-Durkheim wrote in his preface, "but I hesitate to state this for a certainty when I consider the considerable number of organs which I have discovered and described for the first time without having any other work which could serve as my guide." All subsequent works on the skeletal and muscular systems of the cat have been based on Straus-Durckheim's original studies.

By the end of the nineteenth century at least six works on the anatomy of the cat were in print.

The most popular of these was St. George Mivart's *The Cat: An Introduction to the Study of Backboned Animals, Especially Mammals* (1881). Mivart was one of the best-known anatomists of his day. A friend of Thomas Henry Huxley and other prominent Victorian scientists and men of letters, he was for twenty-two years professor of anatomy at St. Mary's Hospital Medical School in London and authored a score of scholarly works on anatomy, natural history, and evolution.

The year 1900 saw the publication of *Anatomy of the Cat,* by Professor Jacob Reighard of the University of Michigan. In his preface Reighard complained that Mivart's book "is written in such general terms that its descriptions are not readily applicable to the actual structures found in the dissection of the cat, and experience has shown that it is not fitted for a laboratory handbook. It contains, in addition to a general account of the anatomy of the cat, a discussion of its embryology, psychology, paleontology, and classification." Reighard set out to remedy these defects by writing a no-nonsense textbook of cat anatomy in the style of *Gray's Anatomy.* His remarkable book, which runs to 585 pages in its third edition, remained in print for over sixty years and is still the standard reference in the field.

In the tradition of his time, Reighard relied on words to convey his message and used small diagrammatic illustrations only to supplement the text. A number of more elaborately illustrated manuals of cat anatomy have been published during the last few decades. By far the best of these is James Crouch's *Text Atlas of Cat Anatomy,* superbly illustrated by Martha B. Lackey (Lea and Febiger, 1969). Although Crouch's *Text Atlas* contains the most detailed and accurate illustrations of cat anatomy in print today, it is too large and too expensive to be purchased by most students in an introductory mammalian anatomy course.

In 1845 Straus-Durckheim complained that one of the greatest problems in comparative anatomy was the lack of a consistent system of names for anatomical structures. A student who compares the names of muscles in the standard references on cat anatomy today may come to the conclusion that not much progress has been made in this department during the last century and a half. For example, the large muscle mass that lies dorsal to the vertebral column is called *extensor dorsi communis* by Reighard, *erector spinae* by Walker and Homburger, *common dorsal extensor of the vertebral column* by Crouch, *sacrospinalis* in the American edition of *Gray's Anatomy,* and *erector spinae* in the British edition of *Gray's Anatomy.*

How did this confusion come about? Reighard, Crouch, and Walker followed three different systems, each of which had been designed with the best of intentions to establish a uniform set of names that would be generally used by anatomists throughout the world. Reighard followed Basle Nomina Anatomica, the anatomical terminology adopted by the German Anatomical Society at their meeting in Basel, Switzerland, in 1895. It seemed like a good idea at the time, but by 1950 it was badly in need of revision and was superseded by *Nomina Anatomica,* which is a list of terms agreed upon by the International Congress of Anatomists for use in human anatomy. Crouch followed the second edition of *Nomina Anatomica,* published in 1961, and many of the terms he uses are therefore different from those used by Reighard. Veterinary anatomists employ yet another system: *Nomina Anatomica Veterinaria,* which is used in textbooks designed for students of veterinary medicine and in some works on comparative anatomy, such as Walker and Homberger's *Vertebrate Dissection.*

The differences in these systems present particularly thorny problems in the names of the muscles. I have tried at least partly to resolve this confusion by listing the equivalent names of some of the principal muscle as found in Reighard, Walker, and Crouch in the table to be found on page 23 of this book.

In *Pictorial Outline of Cat Anatomy With Reference to the Human,* I have attempted to produce a concise laborato.ry manual which is both affordable and practical for use in one-semester courses. In matters of fact and nomenclature I have for the most part followed Reighard and Crouch.

In order to make this book as economical as possible I have not used color in the illustrations, but students who wish to do so should be encouraged to make use of colored pencils as a study aid. Some suggestions: in the illustrations of the skull, bones which develop by intramembranous ossification could be colored yellow and bones derived from the chondrocranium could be colored blue. In illustrations of the muscles, colors could be used to identify muscles with similar actions. Arteries and veins could be colored red and blue, and throughout other systems original color schemes could be devised by the student.

A computer search of current book titles under the subject heading "cat anatomy" turns up an amazing forty-five titles: impressive evidence of the fact that the cat continues to be the specimen of choice as an example of mammalian anatomy.

Suggestions for improving future editions are welcome. Readers can contact me by e-mail at <s.gilbert@utoronto.ca>.

Acknowledgments
I would like to thank my wife, Cheralea, who inspired me, endured me, and collaborated with me throughout the preparation of this book. Dr. Gerry De Iuliis, lecturer in comparative vertebrate anatomy in the Department of Zoology at the University of Toronto, reviewed the text and illustrations and made many valuable suggestions. I am greatly indebted to him for his generous contribution of time and energy. I would also like to thank Dr. Linda Wilson-Pauwels, chair of the Division of Biomedical Communications at the University of Toronto, who provided space for me to work and supported and encouraged me while the work was in progress.

Stephen G. Gilbert
Toronto
November 1999

OUTLINE OF
CAT ANATOMY

with reference to the human

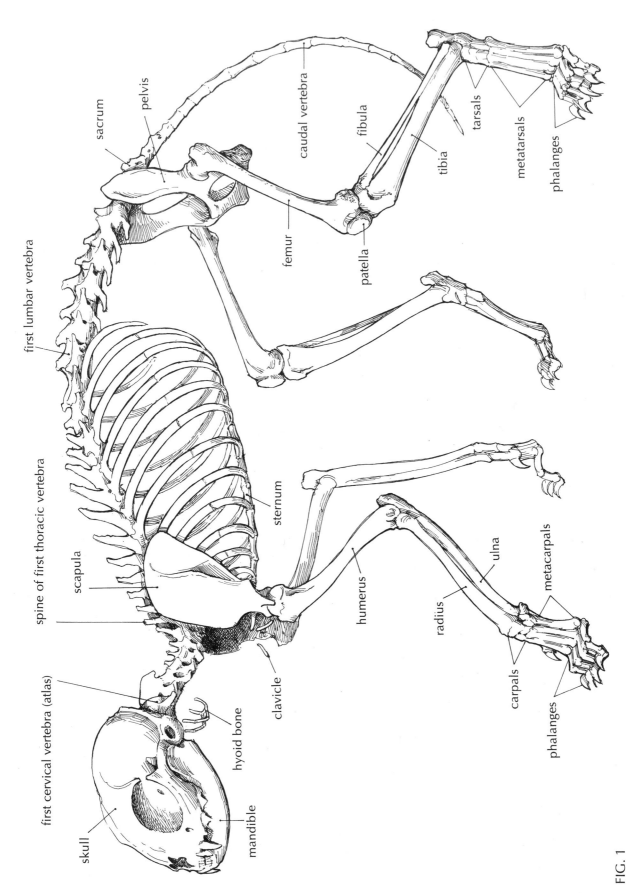

first cervical vertebra (atlas)

skull

mandible

hyoid bone

clavicle

first lumbar vertebra

spine of first thoracic vertebra

scapula

sternum

sacrum

pelvis

caudal vertebra

fibula

tibia

tarsals

metatarsals

phalanges

femur

patella

humerus

radius

ulna

metacarpals

carpals

phalanges

FIG. 1
LATERAL VIEW OF THE SKELETON

THE SKELETON

Articular cartilage. A thin layer of cartilage which covers the articular surfaces of the bones in synovial joints.

Bone marrow. Tissue contained in the cavities of bones. It consists of reticular fibers, fat, and developing blood cells.

Cancellous bone. Spongy bone: a latticework of slender bony spicules.

Cartilage. A specialized fibrous connective tissue that forms most of the temporary skeleton of an embryo and parts of the adult skeletal system.

Compact bone. Dense bone consisting largely of concentric osseous lamellae.

Costal cartilage. A bar of cartilage which attaches a rib to the sternum.

Ligament. A band of fibrous connective tissue which connects two bones.

Medullary cavity. The cavity within the shaft of a bone. It contains bone marrow.

Nutrient foramina. Openings by which arteries, veins, and lymphatics pass through a bone to reach the marrow.

Periosteum. A fibrous membrane which covers the surface of a bone, being absent only at the cartilaginous articulating surfaces.

Bones are connected by the following types of joints:

Movable joints (diarthroses or synovial joints). Includes most joints. The articular surfaces of movable joints are covered by cartilage and lubricated by a clear viscous fluid termed *synovial fluid*. The bones are connected to each other by ligaments lined by *synovial membrane*.

Immovable joints (synarthroses). Example: the sutures between the bones of the skull, in which the bones are held together by interlocking margins united by fibrous tissue.

Slightly movable joints (amphiarthroses). Example: the articulations between the bodies of the vertebrae, in which the vertebral bodies are held together by flattened disks of fibrocartilage termed *intervertebral disks*.

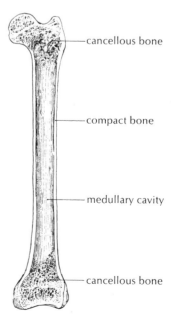

longitudinal section of femur

cancellous bone

compact bone

medullary cavity

cancellous bone

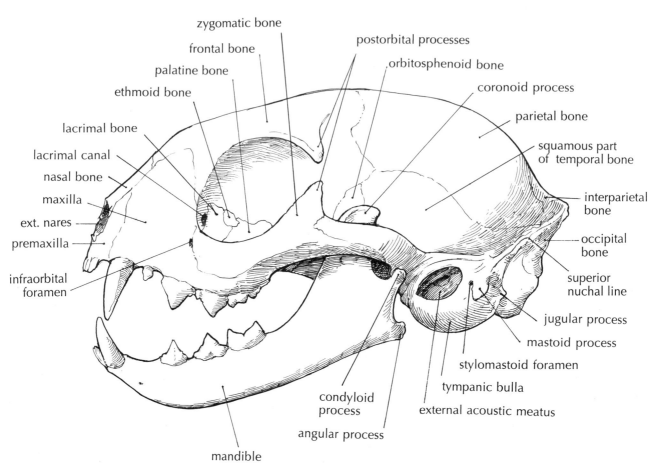

zygomatic bone
frontal bone
palatine bone
ethmoid bone
lacrimal bone
lacrimal canal
nasal bone
maxilla
ext. nares
premaxilla
infraorbital foramen
postorbital processes
orbitosphenoid bone
coronoid process
parietal bone
squamous part of temporal bone
interparietal bone
occipital bone
superior nuchal line
jugular process
mastoid process
stylomastoid foramen
tympanic bulla
external acoustic meatus
condyloid process
angular process
mandible

FIG. 2
LATERAL VIEW OF THE SKULL

Membrane bones. Bones which develop by intramembranous ossification (bone forms directly in mesenchyme).

Membrane bones of the skull:

premaxilla

maxilla

nasal

frontal

parietal

interparietal

squamous part of temporal

vomer

palatine

pterygoid process of basisphenoid

dentary

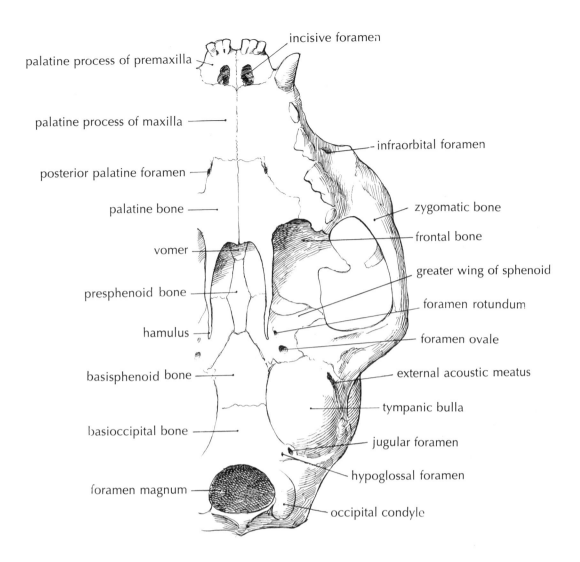

palatine process of premaxilla
incisive foramen
palatine process of maxilla
infraorbital foramen
posterior palatine foramen
zygomatic bone
palatine bone
frontal bone
vomer
greater wing of sphenoid
presphenoid bone
foramen rotundum
hamulus
foramen ovale
basisphenoid bone
external acoustic meatus
tympanic bulla
basioccipital bone
jugular foramen
hypoglossal foramen
foramen magnum
occipital condyle

FIG. 3
BASE OF THE SKULL

Chondrocranium. The cartilaginous base of the embryonic skull. In the adult it is replaced by bone.

Skull bones derived from the chondrocranium:

occipital
petrosal part of temporal
body of basisphenoid
presphenoid
ethmoid
turbinates

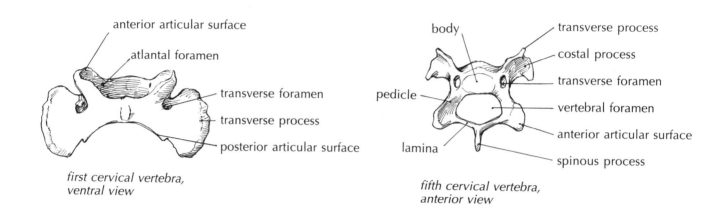

anterior articular surface
atlantal foramen
transverse foramen
transverse process
posterior articular surface

*first cervical vertebra,
ventral view*

body
transverse process
costal process
transverse foramen
pedicle
vertebral foramen
anterior articular surface
lamina
spinous process

*fifth cervical vertebra,
anterior view*

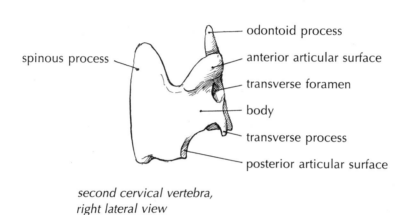

spinous process
odontoid process
anterior articular surface
transverse foramen
body
transverse process
posterior articular surface

*second cervical vertebra,
right lateral view*

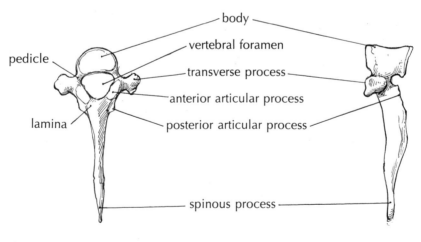

body
pedicle
vertebral foramen
transverse process
anterior articular process
lamina
posterior articular process
spinous process

*fifth thoracic vertebra,
anterior view*

*fifth thoracic vertebra,
right lateral view*

FIG. 4
CERVICAL AND THORACIC VERTEBRAE

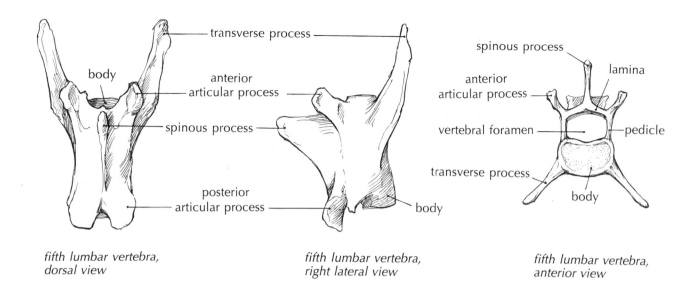

fifth lumbar vertebra,
dorsal view

fifth lumbar vertebra,
right lateral view

fifth lumbar vertebra,
anterior view

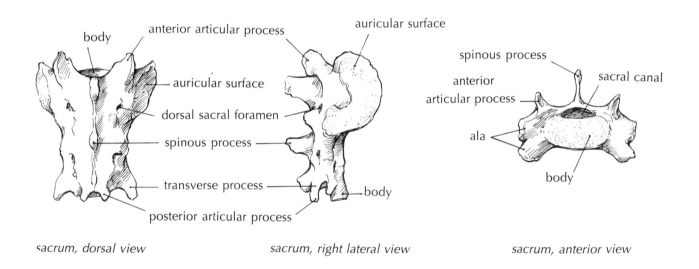

sacrum, dorsal view

sacrum, right lateral view

sacrum, anterior view

FIG. 5
LUMBAR VERTEBRAE AND SACRUM

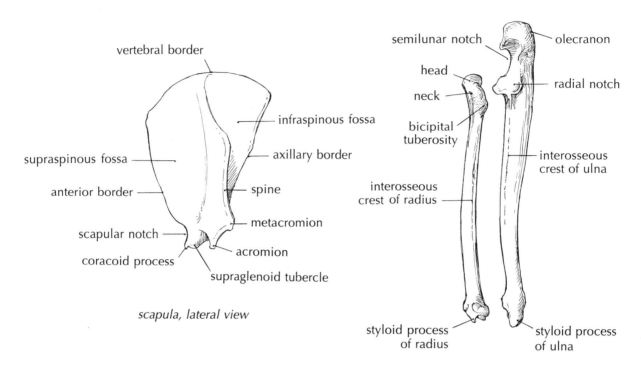

scapula, lateral view

radius and ulna, lateral view

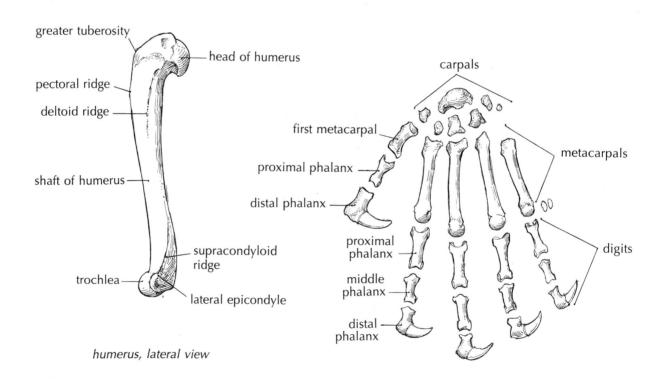

humerus, lateral view

bones of forefoot, dorsal view

FIG. 6
BONES OF THE LEFT FORELIMB

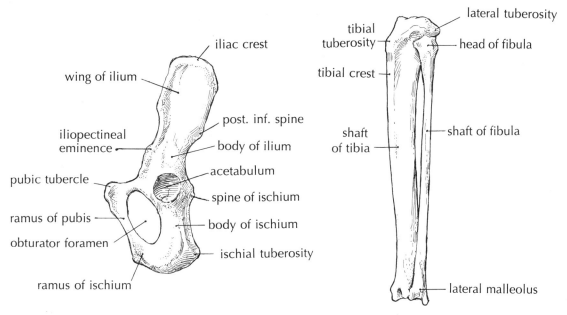

Innominate bone, lateral view

Tibia and fibula labels: lateral tuberosity, tibial tuberosity, head of fibula, tibial crest, shaft of fibula, shaft of tibia, lateral malleolus

Tibia and fibula, lateral view

Innominate bone labels: iliac crest, wing of ilium, post. inf. spine, iliopectineal eminence, body of ilium, pubic tubercle, acetabulum, spine of ischium, ramus of pubis, body of ischium, obturator foramen, ischial tuberosity, ramus of ischium

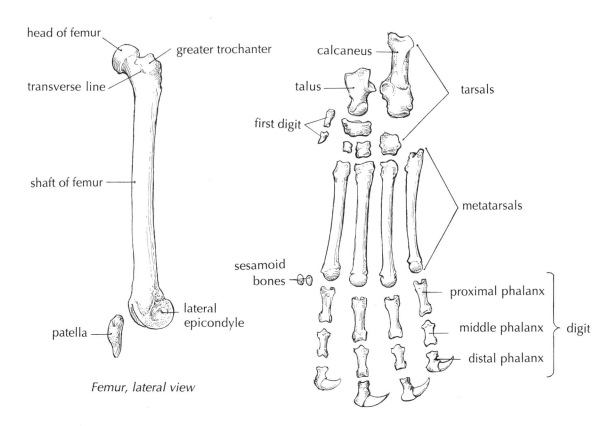

Femur labels: head of femur, greater trochanter, transverse line, shaft of femur, lateral epicondyle, patella

Femur, lateral view

Hindfoot labels: calcaneus, talus, tarsals, first digit, metatarsals, sesamoid bones, proximal phalanx, middle phalanx, distal phalanx, digit

Bones of hindfoot, dorsal view

FIG. 7
BONES OF THE LEFT HINDLIMB

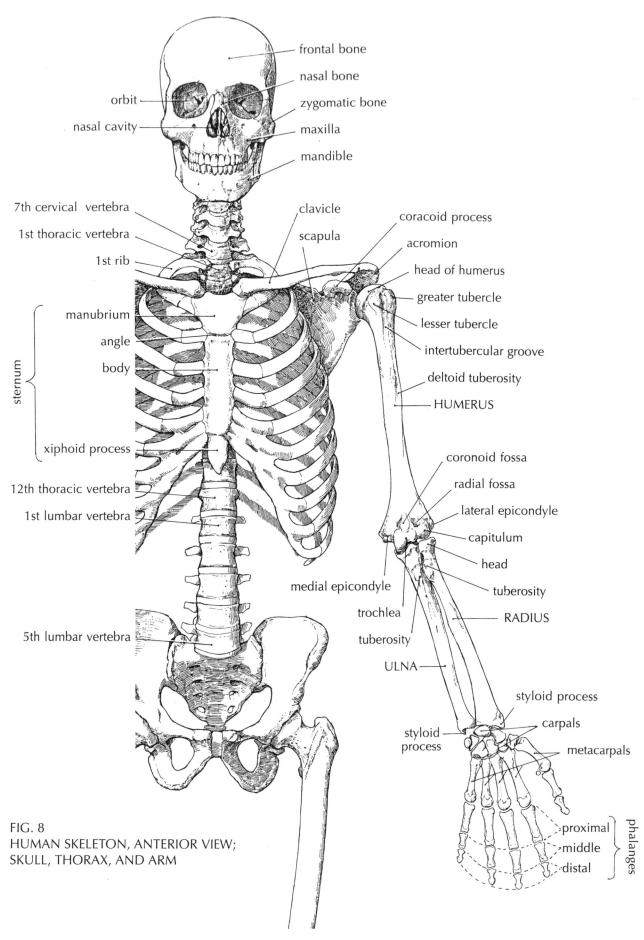

frontal bone

nasal bone

orbit

zygomatic bone

nasal cavity

maxilla

mandible

7th cervical vertebra

1st thoracic vertebra

1st rib

clavicle

scapula

coracoid process

acromion

head of humerus

greater tubercle

lesser tubercle

intertubercular groove

deltoid tuberosity

HUMERUS

manubrium

angle

body

sternum

xiphoid process

12th thoracic vertebra

1st lumbar vertebra

5th lumbar vertebra

coronoid fossa

radial fossa

lateral epicondyle

capitulum

head

tuberosity

medial epicondyle

RADIUS

trochlea

tuberosity

ULNA

styloid process

carpals

styloid process

metacarpals

proximal

middle

distal

phalanges

FIG. 8
HUMAN SKELETON, ANTERIOR VIEW;
SKULL, THORAX, AND ARM

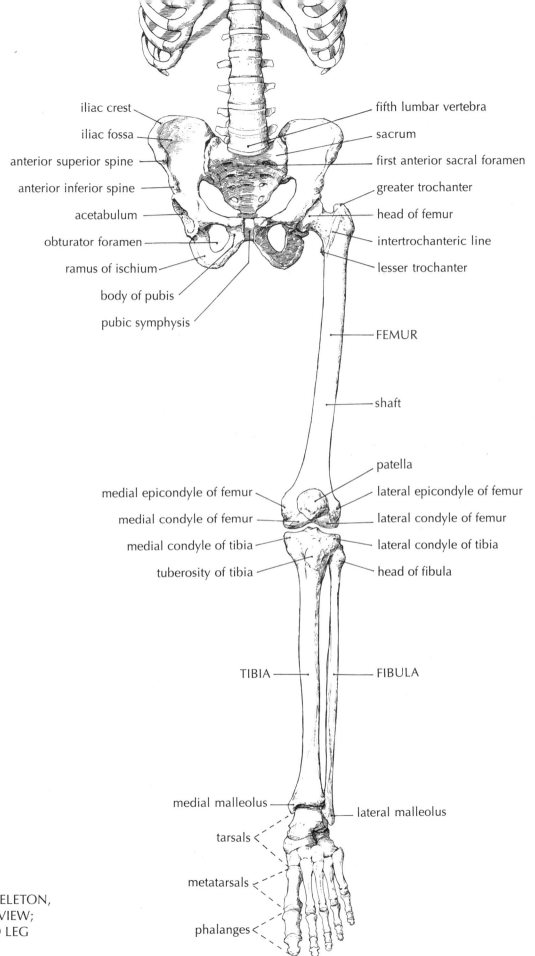

iliac crest

iliac fossa

anterior superior spine

anterior inferior spine

acetabulum

obturator foramen

ramus of ischium

body of pubis

pubic symphysis

fifth lumbar vertebra

sacrum

first anterior sacral foramen

greater trochanter

head of femur

intertrochanteric line

lesser trochanter

FEMUR

shaft

medial epicondyle of femur

medial condyle of femur

medial condyle of tibia

tuberosity of tibia

patella

lateral epicondyle of femur

lateral condyle of femur

lateral condyle of tibia

head of fibula

TIBIA

FIBULA

medial malleolus

lateral malleolus

tarsals <

metatarsals <

phalanges <

FIG. 9
HUMAN SKELETON,
ANTERIOR VIEW;
PELVIS AND LEG

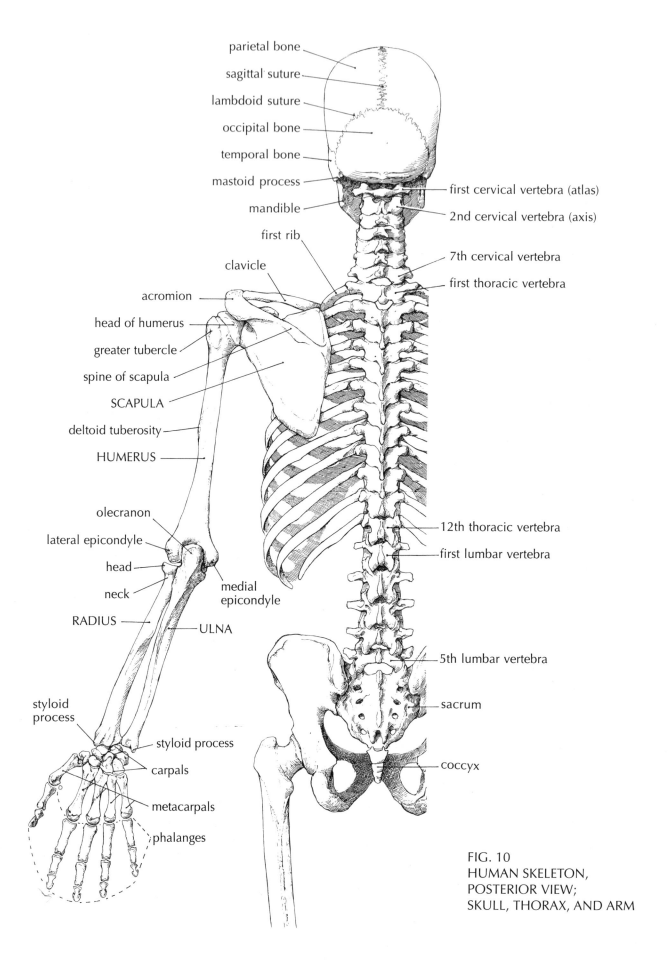

parietal bone
sagittal suture
lambdoid suture
occipital bone
temporal bone
mastoid process
mandible
first rib
clavicle
acromion
head of humerus
greater tubercle
spine of scapula
SCAPULA
deltoid tuberosity
HUMERUS
olecranon
lateral epicondyle
head
neck
RADIUS
medial
epicondyle
ULNA
styloid
process
styloid process
carpals
metacarpals
phalanges

first cervical vertebra (atlas)
2nd cervical vertebra (axis)
7th cervical vertebra
first thoracic vertebra
12th thoracic vertebra
first lumbar vertebra
5th lumbar vertebra
sacrum
coccyx

FIG. 10
HUMAN SKELETON,
POSTERIOR VIEW;
SKULL, THORAX, AND ARM

12

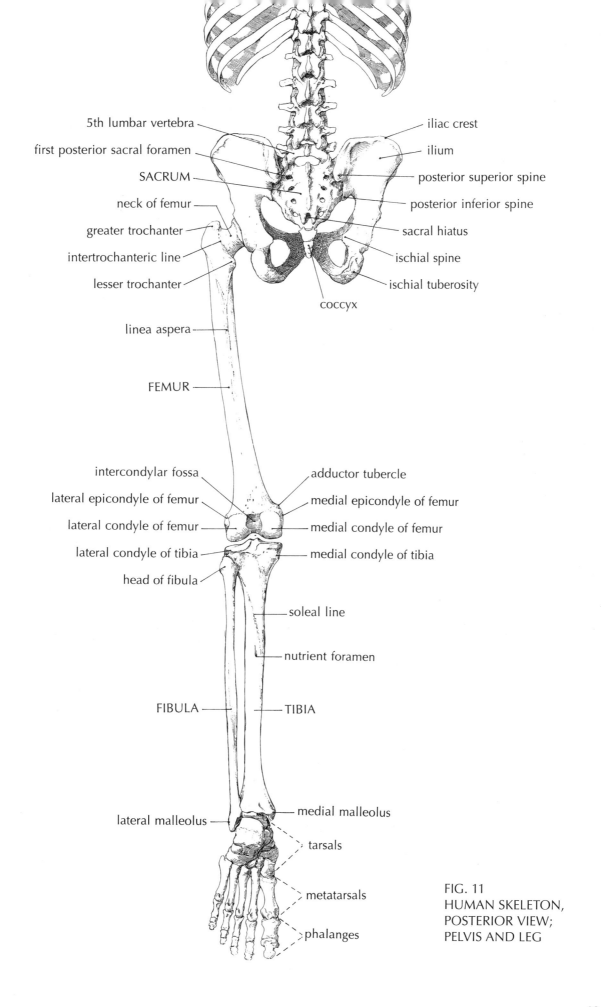

5th lumbar vertebra

first posterior sacral foramen

SACRUM

neck of femur

greater trochanter

intertrochanteric line

lesser trochanter

linea aspera

FEMUR

intercondylar fossa

lateral epicondyle of femur

lateral condyle of femur

lateral condyle of tibia

head of fibula

FIBULA

lateral malleolus

iliac crest

ilium

posterior superior spine

posterior inferior spine

sacral hiatus

ischial spine

ischial tuberosity

coccyx

adductor tubercle

medial epicondyle of femur

medial condyle of femur

medial condyle of tibia

soleal line

nutrient foramen

TIBIA

medial malleolus

tarsals

metatarsals

phalanges

FIG. 11
HUMAN SKELETON,
POSTERIOR VIEW;
PELVIS AND LEG

THE MUSCLES

Separate the skin from the underlying muscles by blunt dissection. Identify two superficial muscles: the *cutaneous maximus*, a large thin muscle which covers most of the side of the body, and the *platysma*, a thin muscle on the sides of the neck and the face. Be careful not to damage superficial veins and glands. When skinning the dorsal side of the thorax and abdomen, observe segmentally arranged cutaneous vessels and nerves.

Abductor. A muscle which moves a limb away from the midline of the body.

Adductor. A muscle which moves a limb toward the midline of the body.

Aponeurosis. A fibrous membrane which encloses a muscle or attaches a muscle to the structure it moves.

Attachment. Muscles may be attached to bones, cartilages, ligaments, fascia, mucous membranes (i.e., tongue muscles), or skin (i.e., facial muscles).

Belly. The fleshy central portion of a voluntary muscle. It consists of contractile muscle fibers and is attached at both ends by connective tissue fibers which form either a tendon or an aponeurosis.

Depressor. A muscle which lowers a structure.

Dilator. A muscle which increases the size of an opening.

Extensor. A muscle which straightens a joint.

Fascia. A sheet of fibrous connective tissue which covers a muscle and/or separates one muscle from another.

Flexor. A muscle which bends a joint.

Insertion. The more mobile end or attachment of a muscle.

Levator. A muscle which raises a structure.

Origin. The more stable end or attachment of a muscle.

Pronator. A muscle which turns the dorsal surface of a limb anteriorly (usually used with forelimbs).

Sphincter. A muscle which closes an opening.

Supinator. A muscle which turns the ventral surface of a limb anteriorly (usually used with hindlimbs).

Tendon. A collagenous band which connects a muscle to a bone.

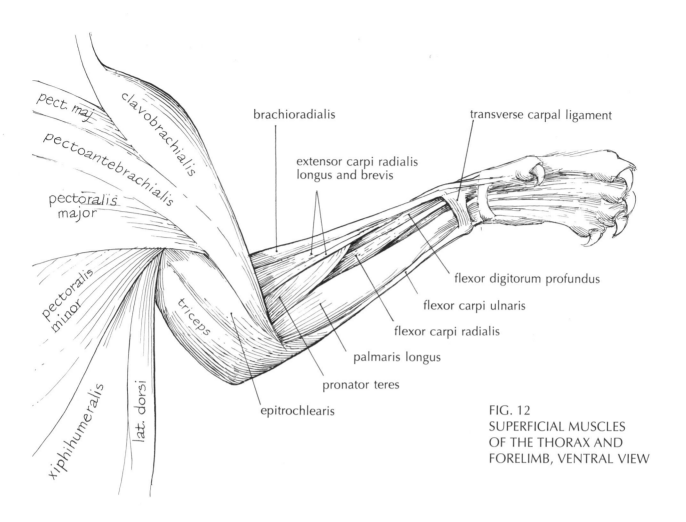

brachioradialis

extensor carpi radialis
longus and brevis

transverse carpal ligament

flexor digitorum profundus

flexor carpi ulnaris

flexor carpi radialis

palmaris longus

pronator teres

epitrochlearis

pect. maj.

clavobrachialis

pectoantebrachialis

pectoralis
major

pectoralis
minor

triceps

xiphihumeralis

lat. dorsi

FIG. 12
SUPERFICIAL MUSCLES
OF THE THORAX AND
FORELIMB, VENTRAL VIEW

Muscle	Origin	Insertion	Action
Clavobrachialis	clavicle; clavotrapezius	ulna	rotates and adducts forelimb
Epitrochlearis	latissimus dorsi	fascia of lower forelimb	extends elbow
Flexor carpi radialis	medial epicondyle of humerus	metacarpals 2 and 3	flexes wrist
Flexor carpi ulnaris	olecranon and medial epicondyle of humerus	carpals	flexes wrist
Palmaris longus	medial epicondyle of humerus	proximal phalanges	flexes wrist and digits
Pecto-antebrachialis	manubrium	fascia of brachium	rotates and adducts forelimb
Pectoralis major and minor	sternum	humerus	rotate and adduct forelimb
Pronator teres	medial epicondyle of humerus	radius	rotates radius
Xiphihumeralis	sternum	humerus	rotates and adducts forelimb

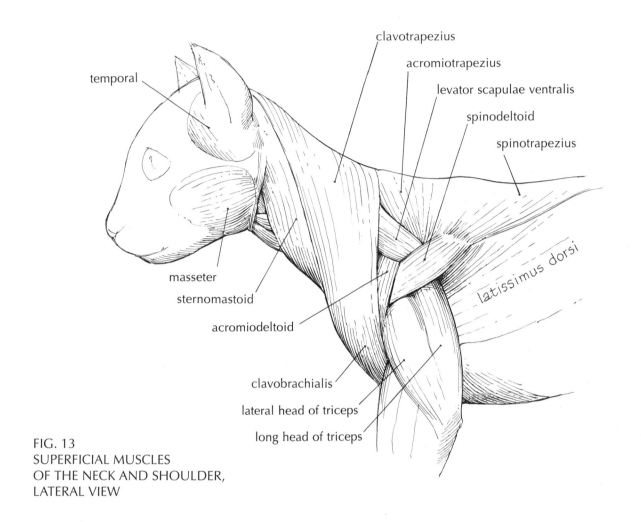

FIG. 13
SUPERFICIAL MUSCLES
OF THE NECK AND SHOULDER,
LATERAL VIEW

Muscle	Origin	Insertion	Action
Acromiodeltoid	acromion	spinodeltoid and humerus	raises and rotates humerus
Acromiotrapezius	spinous processes of cervical and anterior thoracic vertebrae	spine of scapula	pulls scapula toward midline
Clavobrachialis	continuation of clavotrapezius; clavicle	ulna	forward extension of humerus; turns head; extends elbow
Clavotrapezius	sup. nuchal line	clavicle	same as above
Levator scapulae ventralis	occipital bone and atlas	metacromion	draws scapula anteriorly
Masseter	zygomatic bone	dentary bone	elevates mandible
Spinodeltoid	spine of scapula	humerus	acts with acromiodeltoid to raise and rotate humerus
Sternomastoid	manubrium	occipital bone	turns and depresses head
Temporal	temporal fossa	coronoid process of mandible	elevates mandible

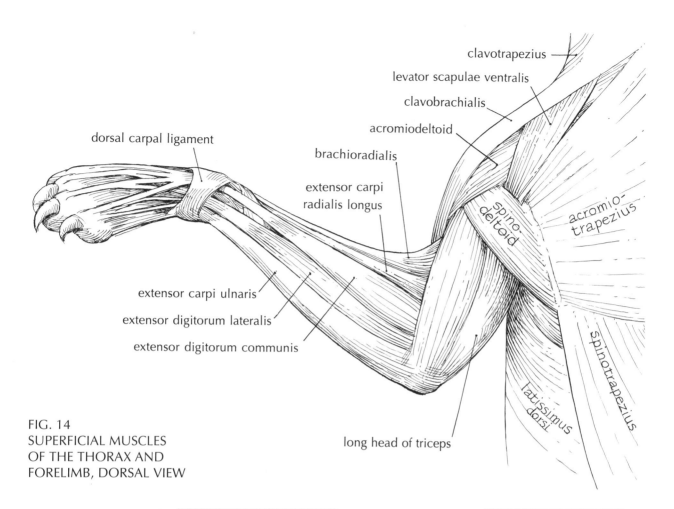

dorsal carpal ligament

clavotrapezius
levator scapulae ventralis
clavobrachialis
acromiodeltoid
brachioradialis
extensor carpi radialis longus
spinodeltoid
acromio-trapezius
spinotrapezius
extensor carpi ulnaris
extensor digitorum lateralis
extensor digitorum communis
latissimus dorsi
long head of triceps

FIG. 14
SUPERFICIAL MUSCLES
OF THE THORAX AND
FORELIMB, DORSAL VIEW

Muscle	Origin	Insertion	Action
Brachioradialis	humerus	radius	rotates radius
Extensor carpi radialis longus and brevis	humerus above lateral epicondyle	2nd and 3rd metacarpals	extend carpal joint
Extensor carpi ulnaris	lateral epicondyle of humerus	5th metacarpal	extends carpal joint
Extensor digitorum communis	humerus above lateral epicondyle	2nd phalanges of digits 2–5	extends digits
Extensor digitorum lateralis	humerus above lateral epicondyle	2nd phalanges of digits 3–5	extends digits
Latissimus dorsi	spinous processes of posterior thoracic and lumbar vertebrae	humerus	pulls forelimb dorsally and posteriorly
Spinotrapezius	spinous processes of posterior thoracic vertebrae	spine of scapula	draws scapula toward midline and posteriorly
Triceps brachii, lateral and medial heads	humerus	olecranon	extend elbow
Triceps brachii, long heads	scapula below glenoid fossa	olecranon	extends elbow

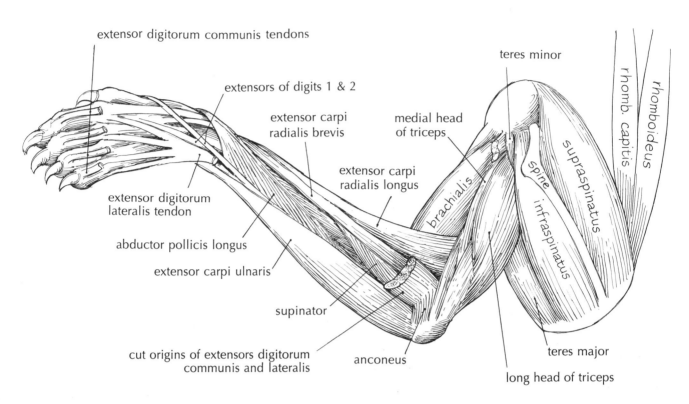

FIG. 15
DEEP MUSCLES OF
THE SHOULDER AND
FORELIMB, DORSAL VIEW

Remove the following muscles: spinotrapezius, acromiotrapezius, levator scapulae ventralis, acromiodeltoid, spinodeltoid, lateral head of triceps, brachioradialis, extensor digitorum communis, and extensor digitorum lateralis. Identify the following muscles:

Muscle	Origin	Insertion	Action
Abductor pollicis longus	ulna	first metacarpal	abducts first digit; extends wrist
Anconeus	humerus	ulna	rotates ulna
Brachialis	humerus	ulna	flexes elbow
Extensors of first and second digits	ulna	1st and 2nd digits	extend lst and 2nd digits
Infraspinatus	infraspinous fossa of scapula	greater tuberosity of humerus	outward rotation of humerus
Supinator	ligaments of elbow	radius	extends wrist supinator of forefoot
Supraspinatus	supraspinous fossa of scapula	greater tubercle of humerus	draws humerus anteriorly
Teres minor	glenoid border of scapula	humerus	outward rotation of humerus

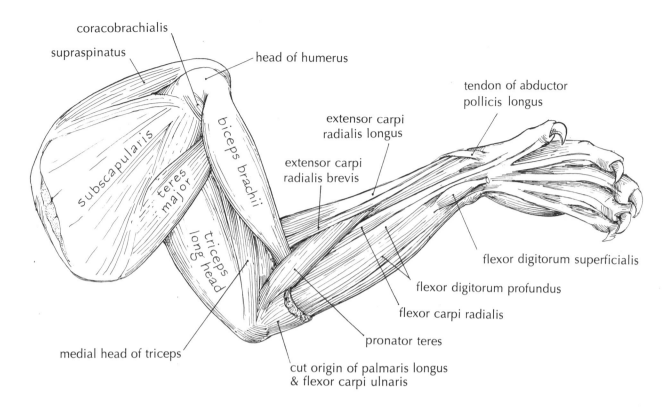

FIG. 16
DEEP MUSCLES OF
THE SHOULDER AND
FORELIMB, VENTRAL VIEW

Cut the rhomboids and the serratus ventralis at their attachments to the scapula.
Remove the palmaris longus, flexor carpi radialis, and flexor carpi ulnaris.

Muscle	Origin	Insertion	Action
Biceps brachii	scapula above glenoid fossa	radius	flexes elbow
Coracobrachialis	coracoid process	humerus	adducts humerus
Flexor digitorum profundus	humerus, radius and ulna	digits 1–5	flexes digits 1–5
Flexor digitorum superficialis	palmaris longus and flexor digitorum profundus	digits 2–5	flexes digits 2–5
Pronator quadratus	ulna	radius	rotates radius
Subscapularis	subscapular fossa	lesser tuberosity of humerus	adducts humerus
Teres major	axillary border of scapula	humerus	rotates humerus and draws it posteriorly

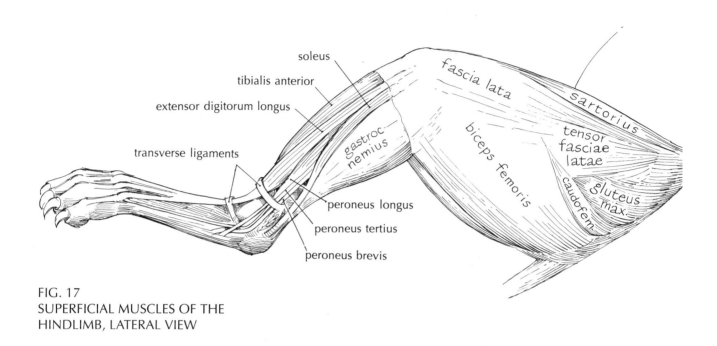

FIG. 17
SUPERFICIAL MUSCLES OF THE
HINDLIMB, LATERAL VIEW

Muscle	Origin	Insertion	Action
Biceps femoris	ischium	patella and tibia	abducts thigh; flexes knee
Extensor digitorum longus	lateral epicondyle of femur	digits 2–5	extends digits
Gastrocnemius	distal end of femur and fascia of knee	calcaneus	extends ankle
Peroneus brevis	fibula	5th metatarsal	extends ankle
Peroneus longus	fibula	metatarsals	flexes ankle
Peroneus tertius	fibula	extensor tendon of 5th digit	flexes ankle; extends 5th digit
Soleus	fibula	calcaneus	extends ankle
Tensor fasciae latae	crest of ilium and nearby fascia	fascia lata	draws thigh forward
Tibialis anterior	tibia and fibula	first metatarsal	flexes ankle

Note: The **fascia lata** is tough white aponeurosis on anterior and lateral aspect of thigh.

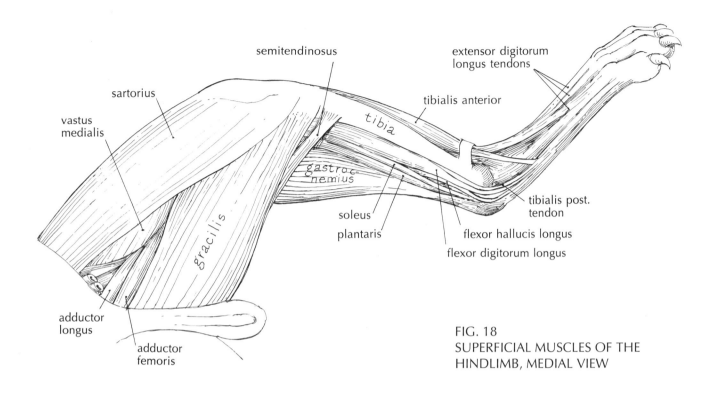

semitendinosus

extensor digitorum longus tendons

sartorius

tibialis anterior

vastus medialis

tibia

gastrocnemius

gracilis

tibialis post. tendon

soleus

plantaris

flexor hallucis longus

flexor digitorum longus

adductor longus

adductor femoris

FIG. 18
SUPERFICIAL MUSCLES OF THE HINDLIMB, MEDIAL VIEW

Muscle	Origin	Insertion	Action
Adductor femoris	pubis	femur	adducts thigh
Adductor longus	pubis	femur	adducts thigh
Flexor digitorum longus	tibia and fibula	phalanges 2–5	flexes digits
Flexor hallucis longus	tibia and fibula	phalanges 2–5	flexes digits
Gracilis	ischium and pubis	tibia	adducts leg and draws it posteriorly
Ilipsoas	lumbar vertebrae	lesser trochanter of femur	rotates and flexes thigh
Pectineus	pubis	femur	adducts thigh
Plantaris	femur and patella	phalanges 2–5	flexes digits and extends ankle
Sartorius	ilium	tibia and patella	adducts leg and draws it posteriorly
Tibialis posterior	tibia and fibula	tarsals	extends ankle

Note: The **vastus medialis** is part of the **quadriceps femoris**, a large muscle situated on the anterior surface of the thigh. It inserts on the patellar ligament and acts as a powerful extensor of the knee.

MUSCLES OF THE VERTEBRAL COLUMN

The **extensor dorsi communis** is a longitudinal muscle mass extending from the sacral region to the head on the dorsal aspect of the vertebral column. Subdivisions of this mass are connected to the pelvis, the vertebrae, ribs, and skull. They extend the spinal column, draw the ribs posteriorly, and bend the neck and spinal column to one side.

Principal divisions of **extensor dorsi communis**:

Iliocostalis—in thoracic region—connects ribs.

Longissimus dorsi (largest part) extends from the pelvis to the neck, filling most of the region between the spinous and transverse processes of the lumbar and thoracic vertebrae.

Multifidus spinae—The deepest fibers; connect vertebrae to each other.

Spinalis dorsi connects spinous processes of the vertebrae in the thoracic and cervical regions.

ABDOMINAL MUSCLES

Identify:

Muscle	Origin	Insertion	Action
External oblique	ribs 4–13 and lumbo-dorsal fascia	linea alba and pubic tubercle	constricts abdomen

Dissection: Cut through the middle of the external oblique and identify:

Internal oblique	pelvis and lumbo-dorsal fascia	linea alba	constricts abdomen
Rectus abdomins	pubis	sternum and costal cartilages 1 and 2	flexes trunk; constricts abdomen
Transversus	ribs 10-13, lumbar vertebrae; ilium	linea alba	constricts abdomen

Reighard	Walker and Homburger	Crouch (if different from Reighard)
acromiotrapezius	cervical trapezius	
clavobrachialis	cleidobrachialis	cleidobrachialis
clavotrapezius	cleidocervicalis	cleidotrapezius
cutaneus maximus	cutaneus trunci	
epitrochlearis	tensor fasciae antebrachii	
extensor brevis pollicis	abductor pollicis longus	abductor pollicis longus
extensor dorsi communis	erector spinae	common dorsal extensor
extensor indicis proprius	extensor digiti I and digiti II	extensor pollicis longus and indicis proprius
flexor sublimis digitorum	flexor digitorum superficialis	flexor digitorum superficialis
palmaris longus	flexor digitorum superficialis (superficial head)	
pectoantebrachialis	pectoralis descendens	
pectoralis major	pectoralis transversus	
pectoralis minor	pectoralis profundus	
rhomboideus	rhomboideus cervicis and thoracis	rhomboideus major and minor
rhomboideus capitis	rhomboideus capitis	occipitoscapularis
serratus anterior	serratus ventralis	serratus ventralis
serratus dorsalis caudalis	serratus posterior superior	
serratus posterior inferior	serratus dorsalis cranials	serratus dorsalis cranialis
serratus posterior superior	serratus dorsalis caudalis	serratus dorsalis caudalis
spinodeltoid	scapulodeltoid	
spinotrapezius	thoracic trapezius	
transversus costarum	rectus thoracis	

References: Crouch 1969; Reighard and Jennings 1935; Walker and Homburger 1992.

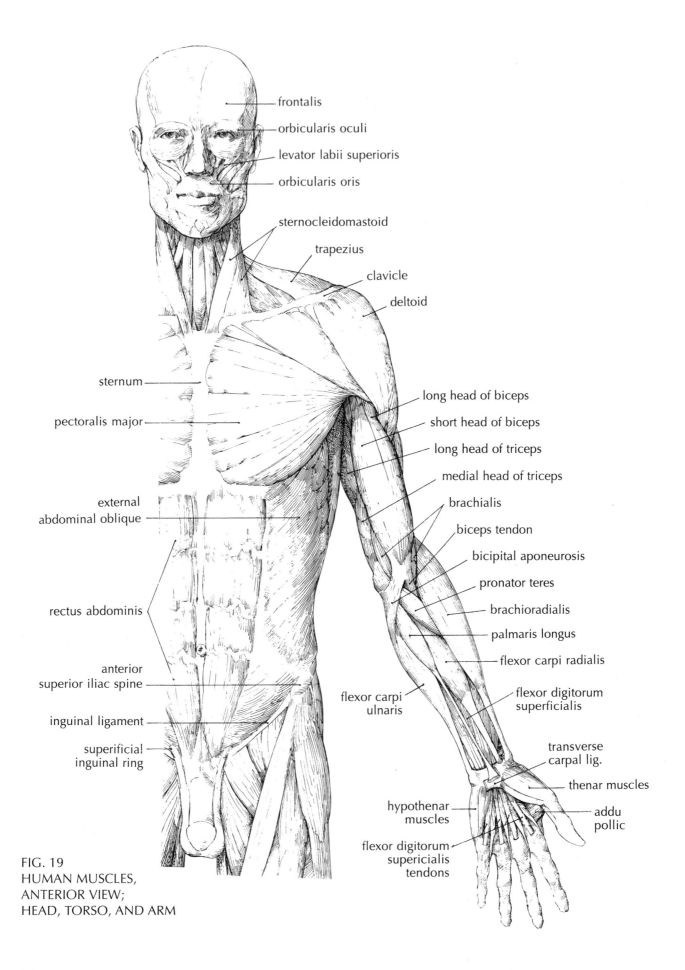

frontalis

orbicularis oculi

levator labii superioris

orbicularis oris

sternocleidomastoid

trapezius

clavicle

deltoid

sternum

long head of biceps

short head of biceps

long head of triceps

medial head of triceps

brachialis

biceps tendon

bicipital aponeurosis

pronator teres

brachioradialis

palmaris longus

flexor carpi radialis

pectoralis major

external
abdominal oblique

rectus abdominis

anterior
superior iliac spine

inguinal ligament

superificial
inguinal ring

flexor carpi
ulnaris

flexor digitorum
superficialis

transverse
carpal lig.

thenar muscles

hypothenar
muscles

addu
pollic

flexor digitorum
supericialis
tendons

FIG. 19
HUMAN MUSCLES,
ANTERIOR VIEW;
HEAD, TORSO, AND ARM

24

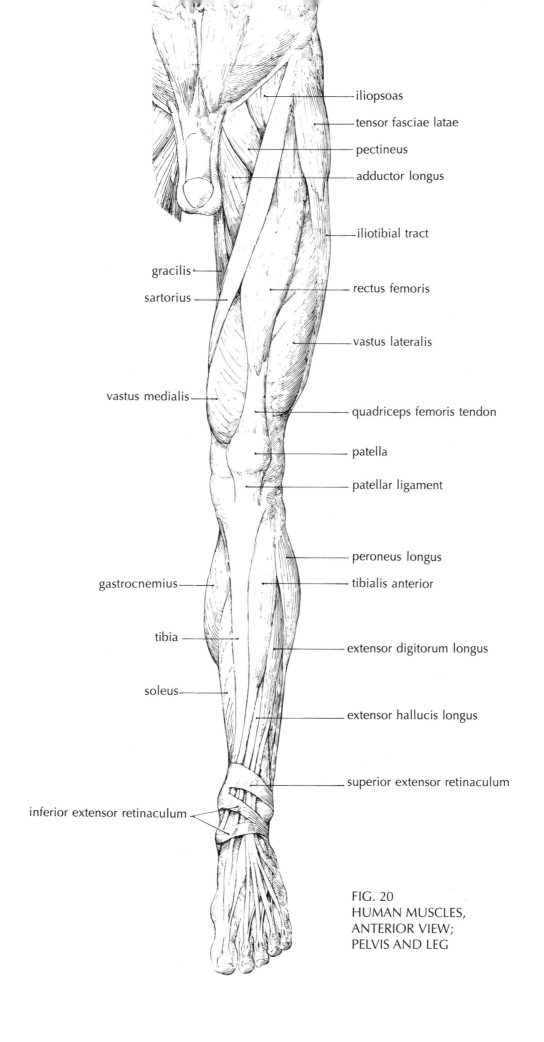

iliopsoas

tensor fasciae latae

pectineus

adductor longus

iliotibial tract

gracilis

sartorius

rectus femoris

vastus lateralis

vastus medialis

quadriceps femoris tendon

patella

patellar ligament

peroneus longus

gastrocnemius

tibialis anterior

tibia

extensor digitorum longus

soleus

extensor hallucis longus

superior extensor retinaculum

inferior extensor retinaculum

FIG. 20
HUMAN MUSCLES,
ANTERIOR VIEW;
PELVIS AND LEG

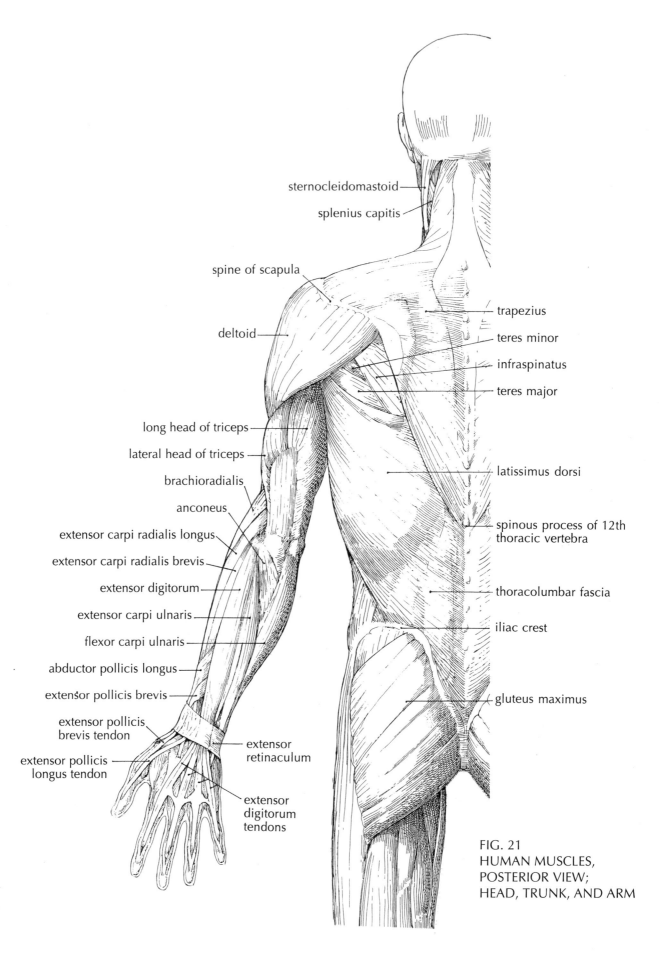

sternocleidomastoid

splenius capitis

spine of scapula

deltoid

trapezius

teres minor

infraspinatus

teres major

long head of triceps

lateral head of triceps

brachioradialis

anconeus

extensor carpi radialis longus

extensor carpi radialis brevis

extensor digitorum

extensor carpi ulnaris

flexor carpi ulnaris

abductor pollicis longus

extensor pollicis brevis

extensor pollicis brevis tendon

extensor pollicis longus tendon

extensor retinaculum

extensor digitorum tendons

latissimus dorsi

spinous process of 12th thoracic vertebra

thoracolumbar fascia

iliac crest

gluteus maximus

FIG. 21
HUMAN MUSCLES,
POSTERIOR VIEW;
HEAD, TRUNK, AND ARM

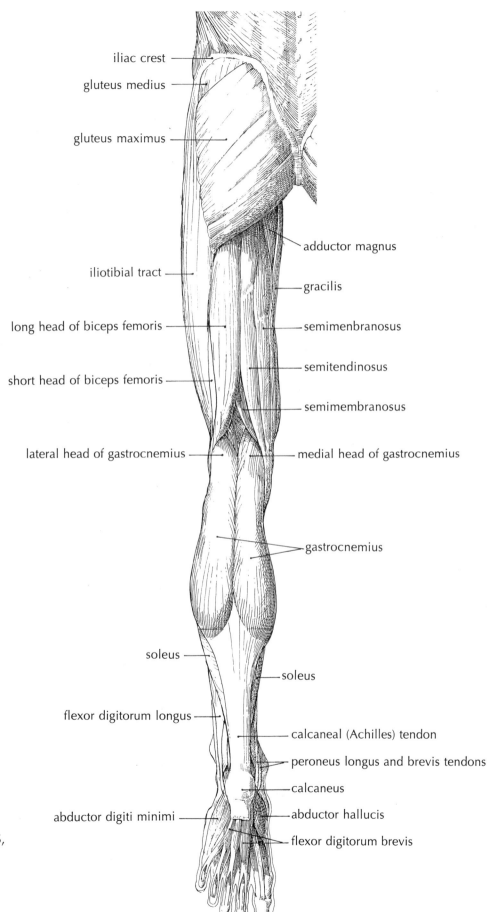

iliac crest

gluteus medius

gluteus maximus

adductor magnus

iliotibial tract

gracilis

long head of biceps femoris

semimenbranosus

semitendinosus

short head of biceps femoris

semimembranosus

lateral head of gastrocnemius

medial head of gastrocnemius

gastrocnemius

soleus

soleus

flexor digitorum longus

calcaneal (Achilles) tendon

peroneus longus and brevis tendons

calcaneus

abductor digiti minimi

abductor hallucis

flexor digitorum brevis

FIG. 22
HUMAN MUSCLES,
POSTERIOR VIEW;
PELVIS AND LEG

THE RESPIRATORY AND DIGESTIVE SYSTEMS

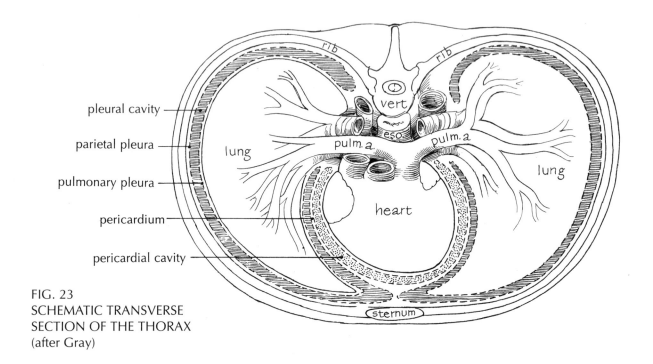

pleural cavity —
parietal pleura —
pulmonary pleura —
pericardium —
pericardial cavity —

FIG. 23
SCHEMATIC TRANSVERSE
SECTION OF THE THORAX
(after Gray)

THE CONTENTS OF THE THORACIC CAVITY

Use a scalpel and bone clippers to cut through the muscles and costal cartilages on either side of the sternum. Extend the cuts from the region of the diaphragm to the first rib. Then, following the space between adjacent ribs, cut through the intercostal muscles on either side about the middle of the thorax. Cut the ribs near their attachment to the vertebral column and pull them back to expose the heart and lungs. Cut into the substance of a lung and observe its spongy texture.

Identify the following structures:

Bronchi. The trachea divides to form right and left main bronchi. Each bronchus subdivides, forming successively smaller branches which terminate in *alveoli* (minute air sacks too small to be seen with the naked eye).

Lungs. The paired organs of respiration. They occupy the right and left halves of the thoracic cavity and are separated by the mediastinum. In the cat, the right

lung consists of four lobes and the left lung consists of three lobes. In the human, the right lung consists of three lobes and the left lung consists of two lobes.

Mediastinum. The space between the two pleural cavities. It contains the heart and all other thoracic viscera except the lungs.

Pleura. The thin serous membrane which invests the lungs and lines the thoracic cavity.

The *pulmonary pleura* covers the lungs.

The *parietal pleura* lines the walls of the thoracic cavity and the cranial surface of the diaphragm.

Pleural cavity. The potential space between the parietal pleura and the pulmonary pleura.

Pulmonary ligament. The pleural fold which attaches the lungs to the aorta, vertebral column, and diaphragm.

Trachea. A tubular air passage composed of cartilages, connective tissue, and mucosa. It extends from the larynx to the bronchii.

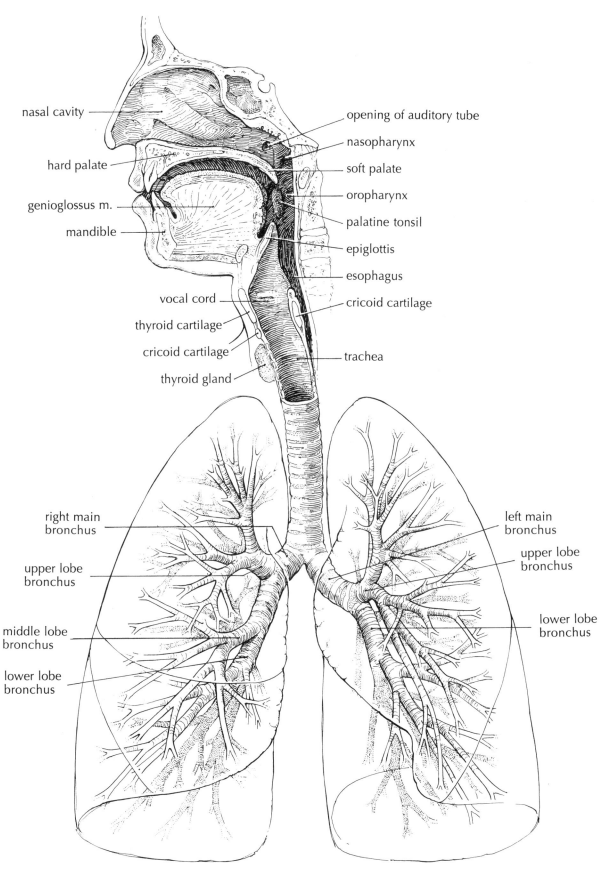

nasal cavity

opening of auditory tube

nasopharynx

hard palate

soft palate

oropharynx

genioglossus m.

palatine tonsil

mandible

epiglottis

esophagus

vocal cord

cricoid cartilage

thyroid cartilage

cricoid cartilage

trachea

thyroid gland

right main
bronchus

left main
bronchus

upper lobe
bronchus

upper lobe
bronchus

middle lobe
bronchus

lower lobe
bronchus

lower lobe
bronchus

FIG. 24
THE HUMAN RESPIRATORY SYSTEM

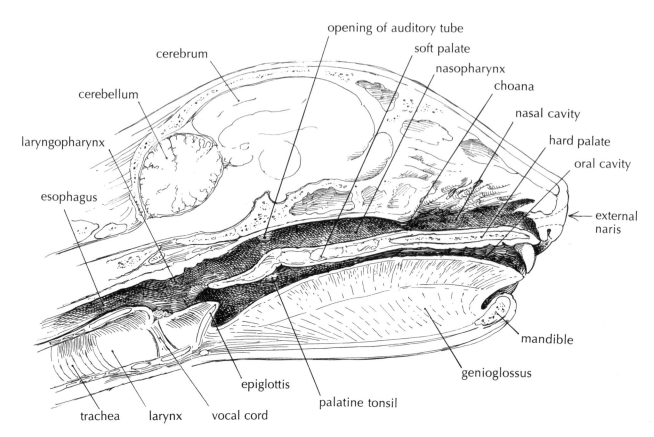

FIG. 25
SAGITTAL SECTION OF THE HEAD AND NECK

Examine a demonstration dissection of a sagittal section of the head and neck. Identify the following structures:

Epiglottis. The cartilage which closes the entrance to larynx during swallowing.

Genioglossus. The largest muscle of the tongue.

Glottis. The vocal cords and the opening between them.

Hard palate. The partition which forms the floor of the nasal cavity and the anterior part of the roof of the oral cavity.

Larynx. A structure consisting of cartilages, ligaments, muscles, and membranes. It lies at the anterior end of the trachea, forming a passageway between the trachea and the pharynx. The cartilages of the larynx are termed the *epiglottic, thyroid, cricoid,* and *arytenoid.*

Nasal cavity. The proximal portion of the respiratory system. Anteriorly it opens via the *external nares;*

posteriorly it communicates with the nasopharynx via the *choanae.* It is divided into symmetrical halves by a cartilaginous partition termed the *median nasal septum.*

Oral cavity. The cavity which extends from the lips to the pharynx.

Palatine tonsil. A small mass of lymphatic tissue which is situated in the lateral mucosal fold on either side near the base of the tongue.

Pharynx. The cavity dorsal to the soft palate and the larynx. Anteriorly it communicates with the nasal cavities; posteriorly it communicates with the oral cavity, larynx, and esophagus. The pharynx consists of two parts: the *nasopharynx,* which lies above the soft palate, and the *laryngopharynx,* which lies above the larynx.

Soft palate. The partition which forms the posterior part of the roof of the oral cavity.

Vocal cords. The mucosal folds which extend between the arytenoid and thyroid cartilages on either side.

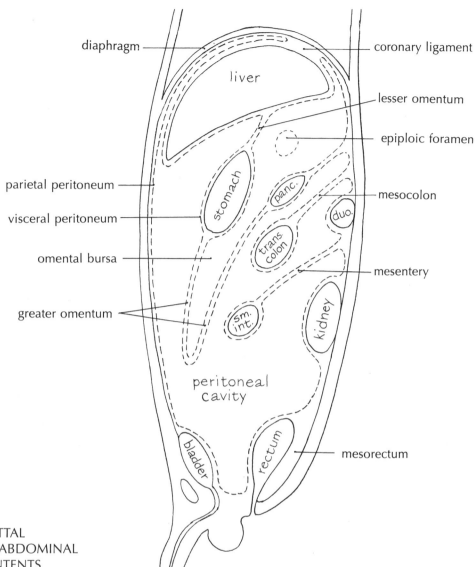

diaphragm —
coronary ligament

liver

lesser omentum

epiploic foramen

parietal peritoneum —

stomach

panc.

mesocolon

visceral peritoneum —

duo.

omental bursa —

trans. colon

mesentery

greater omentum —

kidney

sm. int.

peritoneal cavity

bladder

rectum

mesorectum

FIG. 26
SCHEMATIC SAGITTAL
SECTION OF THE ABDOMINAL
CAVITY AND CONTENTS

Abdominal or peritoneal cavity. The potential space between the parietal peritoneum and the visceral peritoneum.

Diaphragm. A dome-shaped sheet of muscle and tendon which separates the thoracic and abdominal cavities. In the diaphragm are three openings: the *aortic hiatus,* the *esophageal hiatus,* and the *postcaval foramen,* which allow for the passage of the aorta, the esophagus, and the postcava, respectively, between the thorax and the abdomen.

Greater omentum. A double fold of peritoneum attached to the greater curvature of the stomach and to the dorsal wall of the peritoneal cavity.

Lesser omentum. A fold of peritoneum extending from the liver to the duodenum and the lesser curvature of the stomach.

Mesentery. A peritoneal fold which attaches an organ to the body wall.

Mesocolon. The mesentery which supports the colon.

Mesoduodenum. The mesentery which supports the duodenum.

Peritoneum. A serous connective tissue membrane lining the inner body wall *(parietal peritoneum)* and the viscera *(visceral peritoneum).*

Retroperitoneal. Viscera which lie outside or behind (i.e., dorsal to) the peritoneal cavity are said to be *retroperitoneal* (examples: kidneys, bladder, aorta).

Caecum. A blind pouch formed by the caudal end of the colon.

Colon. The first part of the large intestine, consisting of the *caecum* and *ascending, transverse,* and *descending* parts.

Common bile duct. The duct formed by the union of the cystic and hepatic ducts. It joins the pancreatic duct and opens into the duodenum about three centimeters caudal to the pylorus.

Cystic duct. The duct which conveys bile from the gallbladder to the common bile duct.

Duodenum. The proximal U-shaped portion of the small intestine. It encloses the duodenal portion of the pancreas.

Esophagus. The musculo-membranous tube which extends from the pharynx to the stomach. It lies on the ventral aspect of the vertebral column and passes through the esophageal hiatus of the diaphragm.

Gallbladder. The pear-shaped sac which lies on the visceral surface of the right medial lobe of the liver. At its narrow end it is continuous with the cystic duct.

Hepatic ducts. The ducts which convey bile from the liver to the union of the cystic duct and the common bile duct.

Ileocolic junction. The point at which the ileum joins the colon.

Ileocolic valve. A projection of the mucosa and transverse muscle layer around the opening of the ileum into the colon.

Ileum. The distal part of the small intestine which lies between the jejunum and the colon.

Intestine. The musculo-membranous tube which extends from the pylorus to the anus.

Jejunum. The proximal part of the small intestine which lies between the duodenum and the ileum.

Large intestine. That portion of the alimentary canal which extends from the ileocolic junction to the anus. It includes the colon and the rectum.

Liver. The large glandular organ which occupies the cranial part of the abdominal cavity. It is divided into five lobes termed *right medial, right lateral, left medial, left lateral,* and *caudate.* The liver is attached to the diaphragm and to the ventral abdominal wall by ligaments formed by peritoneal folds: the *coronary, triangular,* and *falciform* ligaments, and by the *round ligament* (the remnant of the umbilical vein). See Figure 28.

Pancreas. A flat, lobulated gland consisting of a *gastrosplenic part,* which lies in the dorsal layer of greater omentum, and a *duodenal part,* which lies in the mesoduodenum. *Pancreatic ducts* convey secretions from the pancreas to the duodenum. An *accessory pancreatic duct* opens into the duodenum about 2 cm. posterior to the opening of the common bile duct.

Peyer's patches. Intestinal lymph nodes.

Pyloric outlet. The opening between the stomach and the duodenum.

Pyloric sphincter. The circular layer of muscle fibers surrounding the opening between the stomach and the duodenum. Its position is denoted by the constriction between these structures.

Rectum. The short terminal part of the large intestine which lies in the midline close to the dorsal body wall. It opens externally at the anus.

Small intestine. That part of the intestine which extends from the pylorus to the ileocolic junction. It occupies most of the abdominal cavity and consists of three parts: the duodenum, ileum, and jejunum.

Spleen. The large lymphatic organ which lies between the stomach and the diaphragm on the left side of the abdomen. It is supported by a part of the greater omentum termed the *gastrosplenic ligament.* See Figure 29.

Stomach. The widest part of the digestive tube. It lies at the cranial end of the abdominal cavity, to the left of the midline.

    Greater curvature. The convex caudal margin of the stomach.

    Lesser curvature. The cranial margin of the stomach.

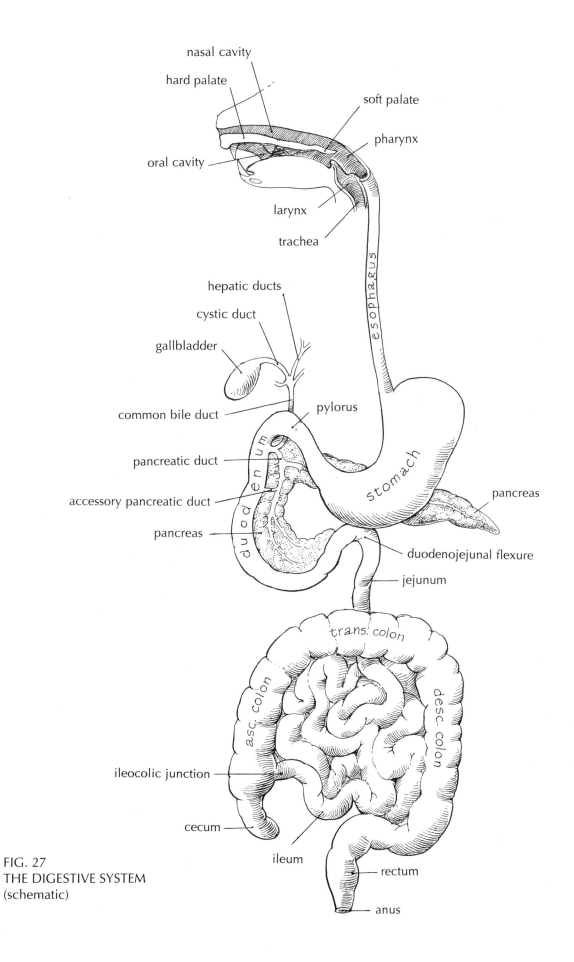

nasal cavity

hard palate

soft palate

oral cavity

pharynx

larynx

trachea

esophagus

hepatic ducts

cystic duct

gallbladder

common bile duct

pylorus

pancreatic duct

duodenum

accessory pancreatic duct

pancreas

stomach

pancreas

duodenojejunal flexure

jejunum

trans. colon

asc. colon

desc. colon

ileocolic junction

cecum

ileum

rectum

anus

FIG. 27
THE DIGESTIVE SYSTEM
(schematic)

33

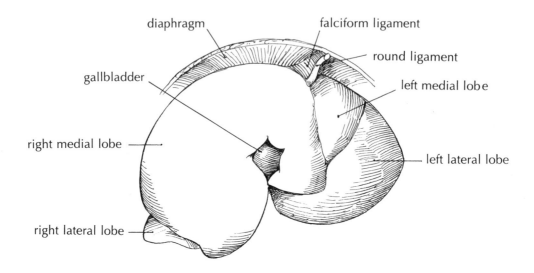

diaphragm falciform ligament

round ligament

gallbladder left medial lobe

right medial lobe left lateral lobe

right lateral lobe

FIG. 28
VENTRAL VIEW OF THE LIVER

Cut the ventral abdominal wall to expose the abdominal viscera. Spread the diaphragm and the liver apart and observe the central tendon of the diaphragm. Muscle fibers originating from the costal cartilages, sternum, vertebrae, and fascia of the dorsal body wall insert on the central tendon.

Identify the lobes of the liver. Lift the liver and find the gallbladder. Dissect the lesser omentum from the caudate lobe of the liver and trace the hepatic, cystic, and common bile ducts. Find the hepatic portal vein and trace it to the liver (see the diagram of the hepatic portal system on page 52). Note that the postcava passes through the substance of the liver, within which it receives the hepatic veins.

Examine the diaphragm and find openings for the passage of the esophagus, postcava, and aorta.

Trim away the greater omentum. Trace the small intestine distal to the pylorus. It passes posteriorly for about three inches and then doubles back on itself, forming a loop within which the duodenal portion of the pancreas lies. Find the common bile duct and the point at which it enters the duodenum. Trace the ileum to its junction with the ascending colon. Identify the caecum and the ascending, transverse, and descending parts of the colon.

Remove the ileum and jejunum, to make a dissection similar to Figure 29. Slit open a section of the jejunum and wash out the contents. Use a hand lens to observe the villi, minute projections which serve to increase the mucosal surface. Identify the celiac artery and its branches. Identify the point at which the hepatic portal vein is formed by the union of the gastrosplenic and superior mesenteric veins.

Remove the colon and the stomach. Open the stomach; observe prominent mucosal folds termed rugae; observe the pyloric sphincter.

Cut open the colon and the ileum and observe the ileocolic valve. Note that there are no villi in the colon.

Examine the pancreas. Identify the point at which the common bile duct joins the duodenum. Dissect pancreatic tissue and trace the pancreatic ducts. Open the duodenum and identify the duodenal papilla (the opening of the duct formed by the union of the common bile duct and the pancreatic duct).

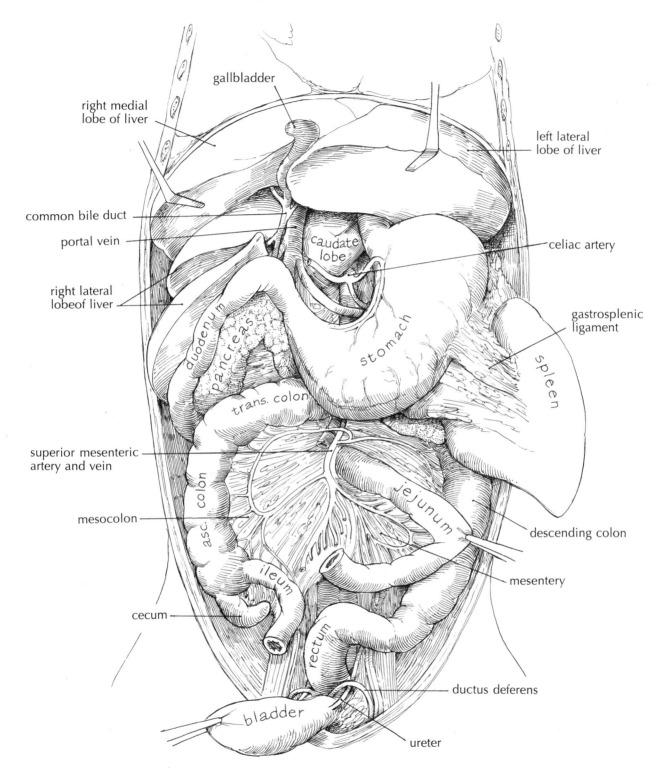

gallbladder

right medial
lobe of liver

left lateral
lobe of liver

common bile duct

caudate
lobe

celiac artery

portal vein

right lateral
lobeof liver

duodenum

pancreas

gastrosplenic
ligament

stomach

spleen

trans. colon

superior mesenteric
artery and vein

jejunum

mesocolon

asc. colon

descending colon

ileum

mesentery

cecum

rectum

ductus deferens

bladder

ureter

FIG. 29
THE ABDOMINAL VISCERA

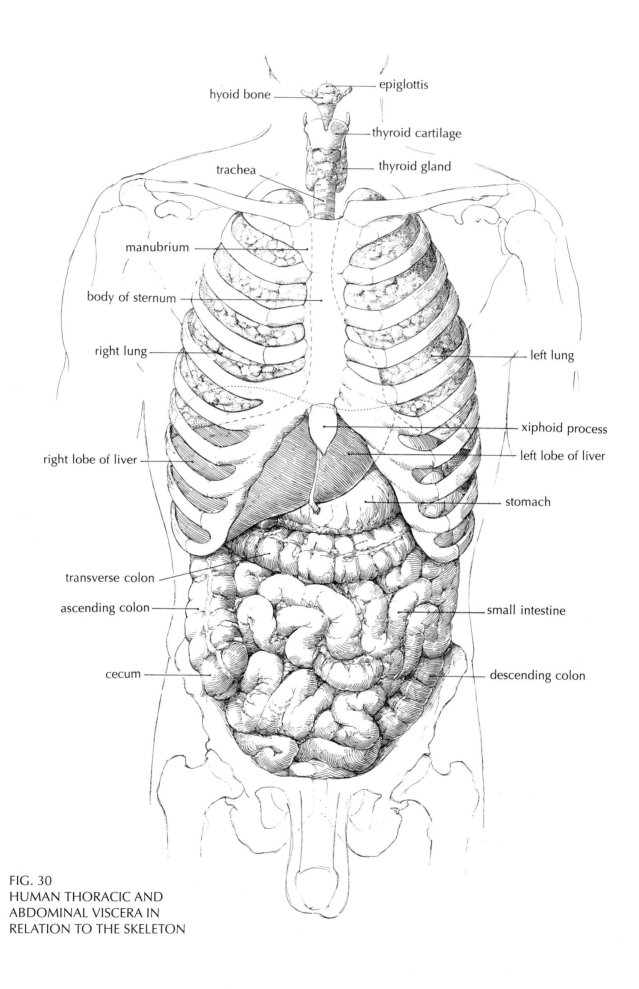

hyoid bone

epiglottis

thyroid cartilage

thyroid gland

trachea

manubrium

body of sternum

right lung

left lung

xiphoid process

left lobe of liver

right lobe of liver

stomach

transverse colon

ascending colon

small intestine

cecum

descending colon

FIG. 30
HUMAN THORACIC AND
ABDOMINAL VISCERA IN
RELATION TO THE SKELETON

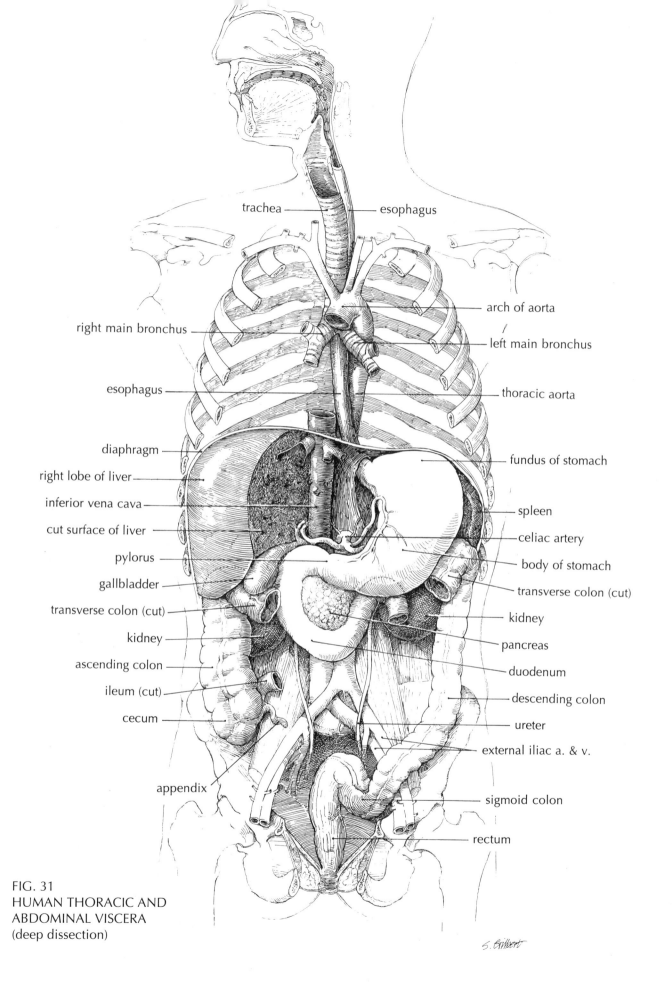

trachea — esophagus

arch of aorta

right main bronchus

left main bronchus

esophagus

thoracic aorta

diaphragm

fundus of stomach

right lobe of liver

inferior vena cava

spleen

cut surface of liver

celiac artery

pylorus

body of stomach

gallbladder

transverse colon (cut)

transverse colon (cut)

kidney

kidney

pancreas

ascending colon

duodenum

ileum (cut)

descending colon

cecum

ureter

external iliac a. & v.

appendix

sigmoid colon

rectum

FIG. 31
HUMAN THORACIC AND
ABDOMINAL VISCERA
(deep dissection)

S. Gilbert

THE UROGENITAL SYSTEM

Remove the liver, spleen, and remaining abdominal viscera. Clear away fat, peritoneum, and connective tissue as necessary to identify the structures illustrated in Figure 32. Remove the fat around the kidneys and strip the renal capsule from the surface of the kidney, observing the attachment of the capsule to the renal vessels and the ureter.

Use a scalpel to make a transverse section of one of the kidneys. Identify the *renal sinus, renal pelvis,* and *renal papilla.* Follow the ureters to the bladder, observing (in the male) the relationship of each ureter to the ductus deferens, or (in the female) to the uterine horns. Make an incision in the ventral wall of the bladder and observe the openings of the ureters and the urethra on the inner surface of the bladder.

Hilus. The central depression in the medial surface of the kidney where the blood vessels, nerves, and ureter enter.

Kidneys. Urinary organs situated on either side of the midline in contact with the dorsal body wall.

Lateral ligaments of the bladder. Paired folds passing from the bladder to the dorsal body wall on either side of the rectum.

Median ligament of the bladder. The peritoneal fold which passes from the bladder to the ventral body wall.

Nephrons. Minute tubules which constitute the functional units of the kidney.

Rectovesical pouch. The space between the bladder and the rectum in the male.

Renal capsule. The fibrous connective tissue capsule which encloses the kidney.

Renal cortex. The peripheral substance of the kidney. It contains the renal corpuscles and convoluted tubules.

Renal medulla. The central substance of the kidney. It contains the collecting tubules.

Renal papilla. The apex of the renal pyramid. On the surface of the papilla are openings of *uriniferous collecting tubules* which convey urine from the substance of the kidney to the pelvis.

Renal pelvis. The expansion at the cranial end of the ureter.

Renal pyramid. The conical mass of collecting tubules which open onto the renal papilla. In the cat there is but one renal pyramid. In humans there are about twelve, and the human renal pelvis forms subdivisions termed *calyces,* each of which embraces one or two papillae.

Renal sinus. The central cavity of the kidney. It contains the renal pelvis.

Suprarenal glands. Paired endocrine glands which lie anterior and medial to the kidneys. Also called *adrenal glands.*

Ureter. The duct which conveys urine from the kidney to the urinary bladder.

Urethra. The duct via which urine leaves the bladder and passes to the exterior of the body.

Urinary bladder. The retroperitoneal muscular sac-like structure located ventral to the rectum.

Vesicouterine pouch. The space between the bladder and the uterus.

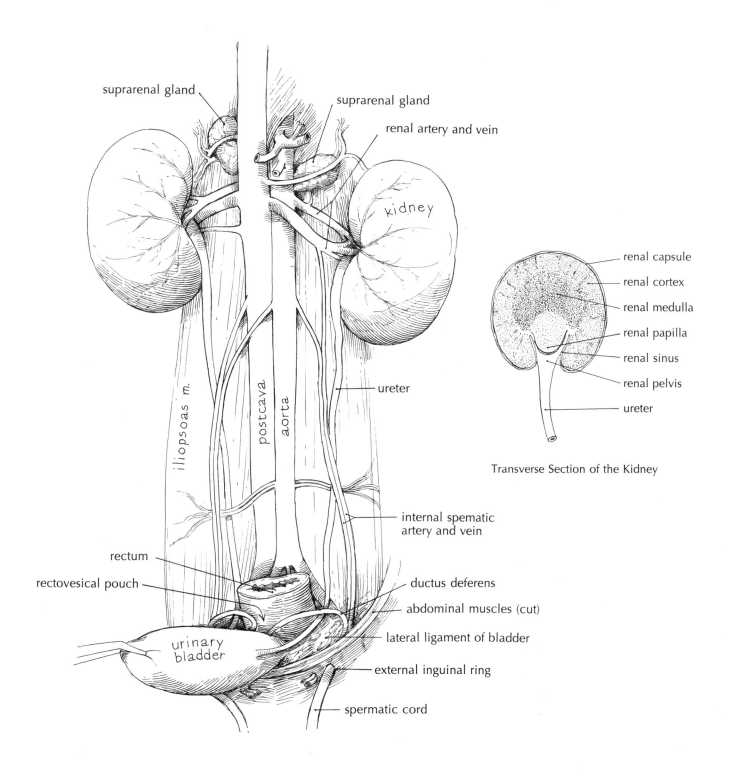

suprarenal gland

suprarenal gland

renal artery and vein

kidney

renal capsule

renal cortex

renal medulla

renal papilla

renal sinus

renal pelvis

ureter

Transverse Section of the Kidney

iliopsoas m.

postcava

aorta

ureter

internal spematic
artery and vein

rectum

rectovesical pouch

ductus deferens

abdominal muscles (cut)

lateral ligament of bladder

urinary
bladder

external inguinal ring

spermatic cord

FIG. 32
THE URINARY ORGANS

Pull back the prepuce and observe the glans, which in the cat is covered with minute horny papillae. Cut open the scrotum and observe that each testis and the spermatic cord are enclosed in a fascial sac. Trace the spermatic cord anteriorly. Observe the external inguinal ring and follow the components of the spermatic cord through the inguinal canal to the internal inguinal ring, identifying the external oblique, internal oblique, and transversus muscles.

Identify the ductus deferens within the abdominal cavity and trace it as it passes dorsal to the ureter. Remove the body of the pubis and the ramus of the ischium. Trace the urethra from the bladder to the penis, observing the prostate gland, the point where the ductus deferens joins the urethra, and the bulbourethral glands. Cut open the fascial sac and observe the testes, epididymis, and ductus deferens. Make a transverse section of the penis and observe the corpora cavernosa and the corpus spongiosum.

Bulbourethral glands. Paired glands near the base of the penis.

Corpora cavernosa. Paired fibrous cylindrical structures within the penis. They contain numerous blood sinuses.

Corpus spongiosum. A cylindrical structure which encloses the penile portion of the urethra.

Crus of penis. The tapering proximal portion of the corpus cavernosum. It is attached on either side to the ramus of the ischium and the pubis.

Ductus deferens. A tube which conveys spermatozoa and components of semen from the epididymis to the prostate gland and urethra.

Epididymis. The first portion of the excretory duct of the testis. It lies on the medial surface of the testis and consists of tubules which convey spermatozoa from the testis to the ductus deferens.

External inguinal ring. The external opening of the inguinal canal.

Glans penis. The enlarged distal end of the corpus spongiosum.

Gubernaculum. A fold of fibrous tissue which connects the inferior pole of the testis to the developing scrotal pouch in the fetus. It functions in the descent of the testis.

Inguinal canal. The opening through which components of the spermatic cord pass through the abdominal wall.

Internal inguinal ring. The internal opening of the inguinal canal.

Prepuce. The protective sheath of skin covering glans.

Prostate gland. A bi-lobed gland on the dorsal aspect of the urethra.

Scrotal ligament. The remnant of the gubernaculum in the adult.

Scrotum. A pouch consisting of skin, fascia, and non-striated muscular fibers. An internal septum divides it into two compartments, each of which contains a testis.

Spermatic cord. A structure consisting of the ductus deferens, nerves, lymphatics, and vessels which supply the testes, enclosed in a layer of muscle and a fascial sheath. The spermatic cord extends from the testis through the inguinal canal to the internal inguinal ring.

Testes. Paired oval organs which produce spermatozoa and components of semen.

Vasa efferentia testis. Tubules within the testes which convey spermatozoa to the epididymis.

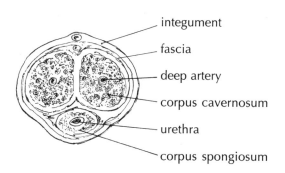

integument
fascia
deep artery
corpus cavernosum
urethra
corpus spongiosum

Transverse Section of the Human Penis (after Gray)

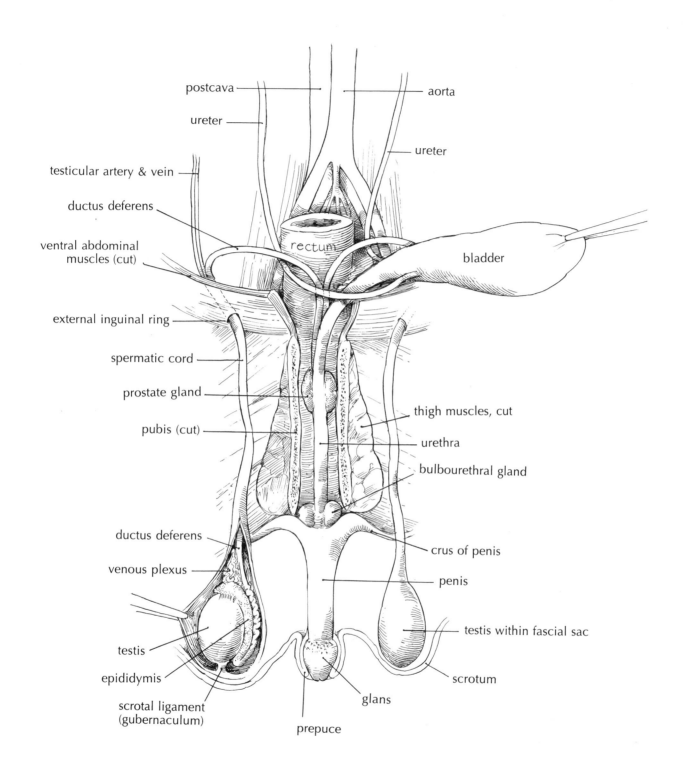

postcava

aorta

ureter

ureter

testicular artery & vein

ductus deferens

ventral abdominal muscles (cut)

rectum

bladder

external inguinal ring

spermatic cord

prostate gland

pubis (cut)

thigh muscles, cut

urethra

bulbourethral gland

ductus deferens

venous plexus

crus of penis

penis

testis

epididymis

testis within fascial sac

scrotum

scrotal ligament (gubernaculum)

glans

prepuce

FIG. 33
THE MALE GENITAL ORGANS

Using bone clippers, remove portions of the pubis and ischium to expose the urethra, the body of the uterus, the vagina, and the urogenital sinus. Identify the structures illustrated in Figure 34. On one side cut open the uterine tube and uterine horn, tracing the lumen as far as the body of the uterus. Introduce a probe through the body of the uterus and the cervix into the vagina. Make a cut to one side of the midline in the urogenital sinus and the vagina. Pass a probe through the bladder and the urethra, and identify the external urethral orifice and the clitoris on the ventral surface of the urogenital sinus.

Abdominal ostium of the uterine tube. The opening through which the ovum enters the uterine tube.

Broad ligament (mesometrium). The peritoneal fold which attaches the uterus and ovary to the dorsal body wall.

Cervix. The portion of the uterus enclosed by the vagina: a small rounded projection formed by the body of the uterus at the uterovaginal junction.

Clitoris. The homolog of the penis. It lies on the ventral aspect of the urogenital aperture.

External urethral orifice. The opening of the urethra into the urogenital sinus.

External uterine orifice. The opening in the cervix via which the lumen of the uterus communicates with the vagina.

Fimbriae. Minute projections surrounding the abdominal ostium of the uterine tube.

Labia. Folds of skin on either side of the urogenital aperture.

Mesovarium. The part of the broad ligament which supports the ovary.

Ovarian ligament. A short fibrous cord which attaches the ovary to the uterine horn.

Ovaries. The female gonads. They lie within the abdominal cavity caudal to the kidneys.

Round ligament. The thin fibrous band which attaches the cranial end of the uterine horn to the body wall.

Urogenital aperture. The external opening of the urogenital sinus.

Urogenital sinus (vestibule). The common passage formed by the union of the urethra and the vagina.

Uterine (Fallopian) tube. The duct through which the ovum passes from the ovary to the uterus.

Uterus. The Y-shaped organ in which the fetus develops. It consists of a *body* and two *horns.*

Vagina. The tubular structure which extends from the cervix to the opening of the urethra.

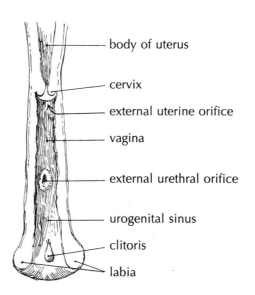

Dorsal View of the Interior of the Vagina and Urogenital Sinus.

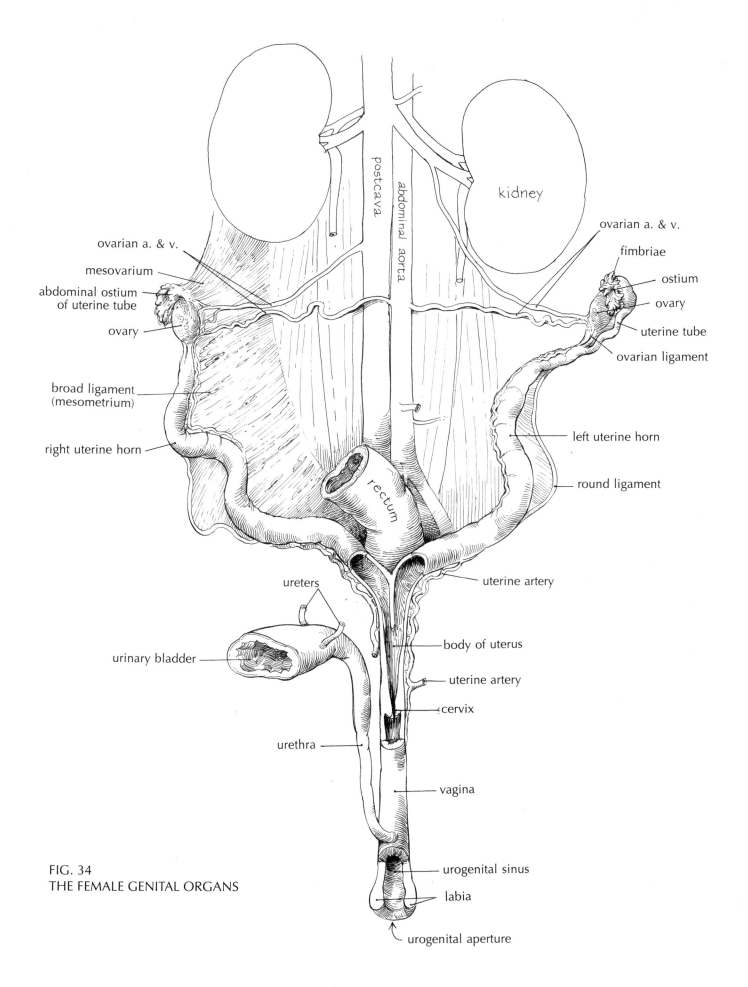

ovarian a. & v.

mesovarium

abdominal ostium
of uterine tube

ovary

broad ligament
(mesometrium)

right uterine horn

postcava

abdominal aorta

kidney

ovarian a. & v.

fimbriae

ostium

ovary

uterine tube

ovarian ligament

left uterine horn

round ligament

rectum

ureters

urinary bladder

urethra

uterine artery

body of uterus

uterine artery

cervix

vagina

urogenital sinus

labia

urogenital aperture

FIG. 34
THE FEMALE GENITAL ORGANS

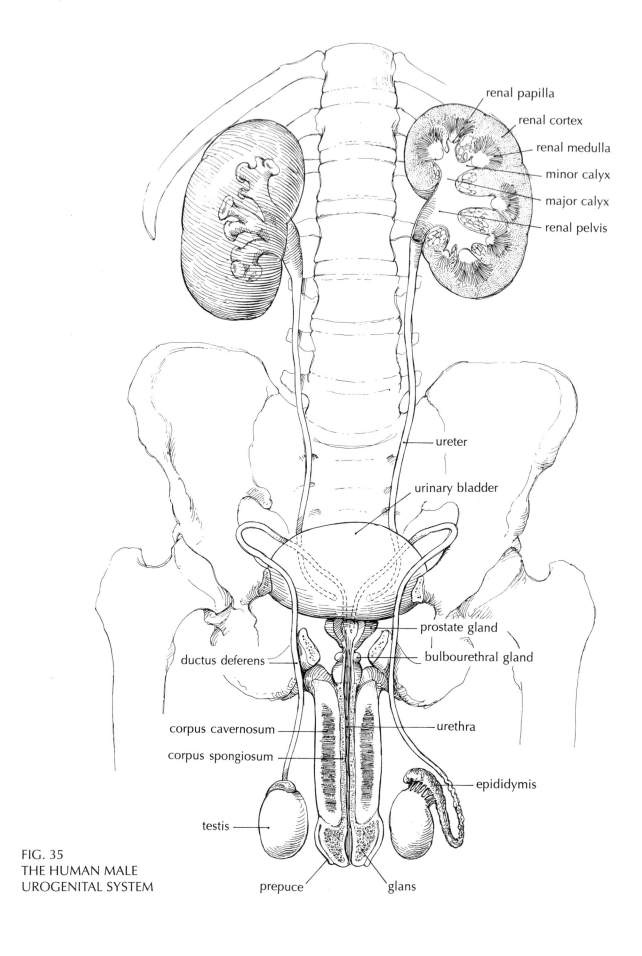

renal papilla

renal cortex

renal medulla

minor calyx

major calyx

renal pelvis

ureter

urinary bladder

prostate gland

ductus deferens

bulbourethral gland

corpus cavernosum

urethra

corpus spongiosum

epididymis

testis

prepuce

glans

FIG. 35
THE HUMAN MALE
UROGENITAL SYSTEM

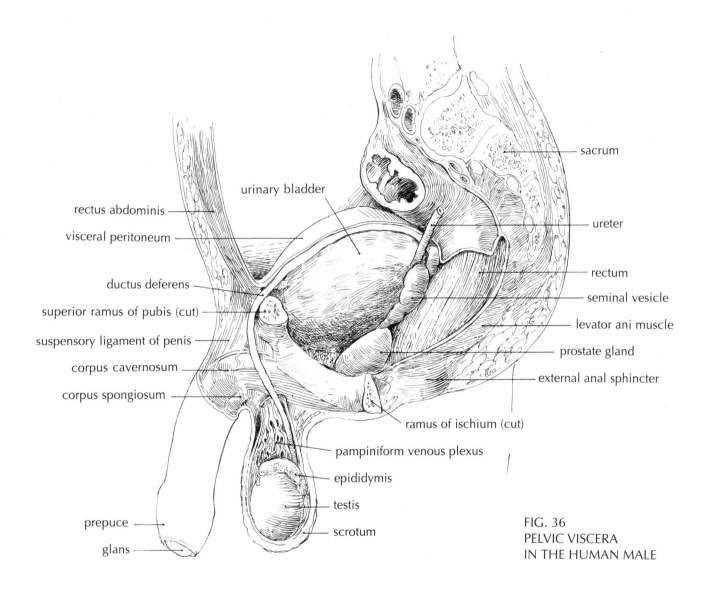

rectus abdominis

visceral peritoneum

urinary bladder

sacrum

ureter

ductus deferens

superior ramus of pubis (cut)

suspensory ligament of penis

corpus cavernosum

corpus spongiosum

rectum

seminal vesicle

levator ani muscle

prostate gland

external anal sphincter

ramus of ischium (cut)

pampiniform venous plexus

epididymis

testis

prepuce

glans

scrotum

FIG. 36
PELVIC VISCERA
IN THE HUMAN MALE

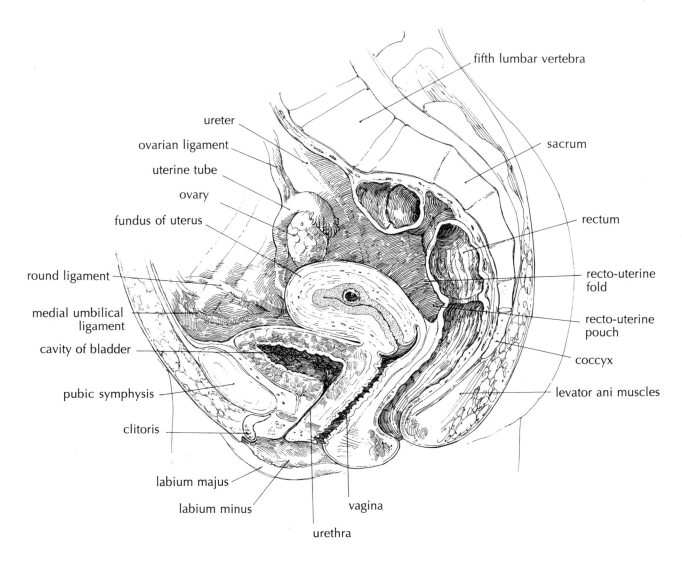

fifth lumbar vertebra

ureter

ovarian ligament

uterine tube

ovary

fundus of uterus

round ligament

medial umbilical ligament

cavity of bladder

pubic symphysis

clitoris

labium majus

labium minus

urethra

vagina

sacrum

rectum

recto-uterine fold

recto-uterine pouch

coccyx

levator ani muscles

FIG. 37
SAGITTAL SECTION OF THE
HUMAN FEMALE PELVIS
A two-week old blastocyst is
implanted in the endometrium.

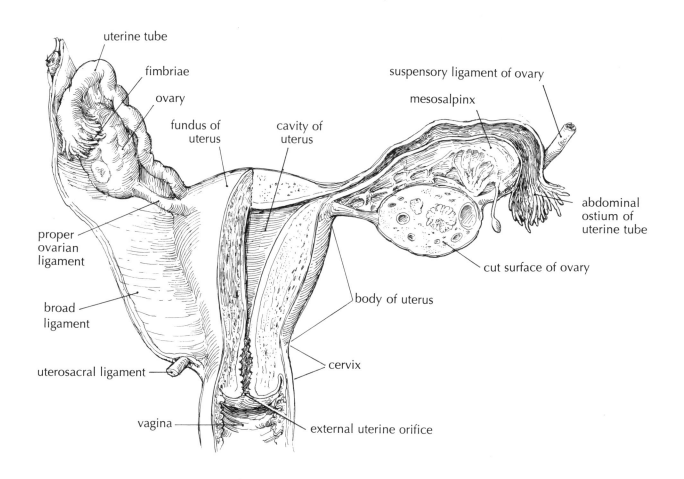

uterine tube

fimbriae

ovary

fundus of
uterus

cavity of
uterus

suspensory ligament of ovary

mesosalpinx

abdominal
ostium of
uterine tube

proper
ovarian
ligament

cut surface of ovary

broad
ligament

body of uterus

uterosacral ligament

cervix

vagina

external uterine orifice

FIG. 38
POSTERIOR VIEW OF
THE HUMAN UTERUS

THE CIRCULATORY SYSTEM

Pull back the lungs to expose the heart. Remove pericardial fat and the *thymus* (a flattened gland-like organ which lies in the mediastinal cavity ventral to the heart). Refer to Figure 23 (page 28) and observe the relationship of the pericardium to the heart and pleura. Make a longitudinal cut in the external fibrous layer of the pericardium, pull it back, and observe the attachments of the fibrous and serous layers to the heart. Trim away the outer layer of the pericardium and identify the structures illustrated in Figure 39, but do not remove the heart at this time. Refer to a demonstration dissection of a heart which has been removed from the thorax and identify the structures illustrated in Figure 40.

Conus arteriosus. The superior part of the right ventricle which leads into the pulmonary trunk.

Coronary arteries. Right and left coronary arteries arise from the aorta near its origin and supply oxygenated blood to the heart.

Coronary sinus. A venous channel on the dorsal surface of the heart. It receives blood from the substance of the heart and returns it to the right atrium.

Ductus arteriosus. A vessel which conveys blood from the pulmonary trunk to the arch of the aorta in the fetus.

Left atrium. The chamber of the heart which receives oxygenated blood from the lungs via the pulmonary veins. It opens into the left ventricle.

Left auricle. A flap-like appendage of the wall of the left atrium.

Left ventricle. The chamber of the heart which receives oxygenated blood from the left atrium and conveys it to the aorta.

Ligamentum arteriosum. A fibrous cord representing the obliterated ductus arteriosus of the fetal heart.

Pericardium. The membranous sac which encloses the heart. It consists of two layers: an *external fibrous layer* and an *internal serous layer.*

Pulmonary arteries. The pulmonary trunk divides into right and left pulmonary arteries. The right pulmonary artery passes dorsal to the ascending aorta and divides into branches which supply the right lung. The left pulmonary artery passes ventral to the descending aorta and divides into branches which supply the left lung.

Pulmonary trunk. A short, wide vessel which conveys unoxygenated blood from the right ventricle to the right and left pulmonary arteries.

Pulmonary veins. The veins which return blood from the lungs to the left atrium.

Right atrium. The chamber of the heart which receives unoxygenated blood from the precava, postcava, and coronary sinus.

Right auricle. A flap-like projection of the wall of the right atrium.

Right ventricle. The chamber of the heart which receives unoxygenated blood from the right atrium. It opens into the pulmonary artery.

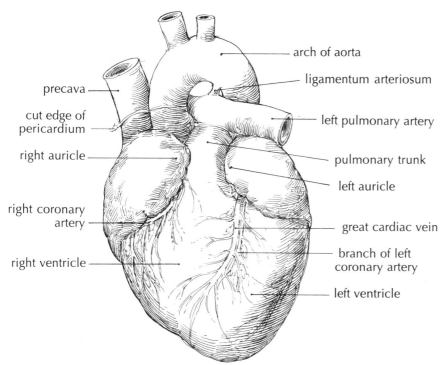

precava

cut edge of
pericardium

right auricle

right coronary
artery

right ventricle

arch of aorta

ligamentum arteriosum

left pulmonary artery

pulmonary trunk

left auricle

great cardiac vein

branch of left
coronary artery

left ventricle

FIG. 39
VENTRAL VIEW OF THE HEART

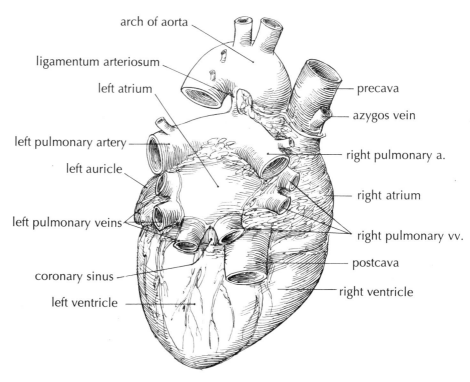

arch of aorta

ligamentum arteriosum

left atrium

left pulmonary artery

left auricle

left pulmonary veins

coronary sinus

left ventricle

precava

azygos vein

right pulmonary a.

right atrium

right pulmonary vv.

postcava

right ventricle

FIG. 40
DORSAL VIEW OF THE HEART

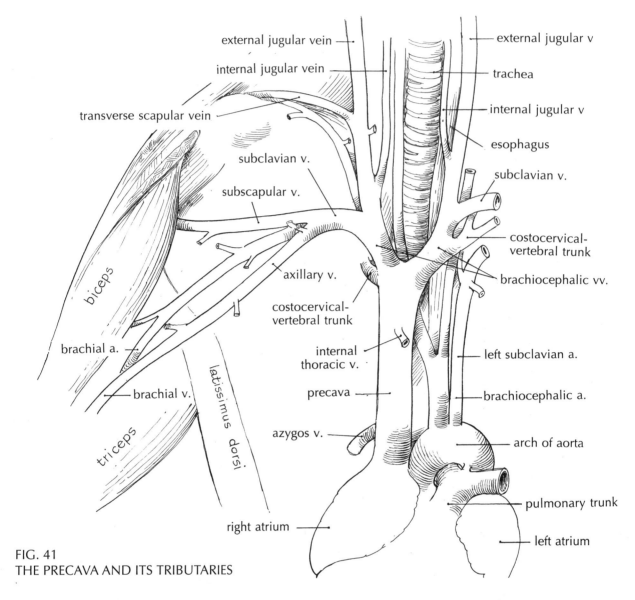

FIG. 41
THE PRECAVA AND ITS TRIBUTARIES

Cut and remove the pectoral muscles. Trim away the thoracic wall and remove fat and connective tissue to expose the vessels and nerves of the forelimb. Identify the structures illustrated in Figure 41.

Azygos vein. Enters precava on the right side just cranial to the root of the right lung. The azygos vein receives the intercostal veins and veins from the muscles of the dorsal abdominal wall, esophagus, and bronchi.

Brachiocephalic veins. Right and left brachiocephalic veins unite to form the precava. Each vein is formed by the union of the external jugular and subclavian veins; each receives a common trunk formed by the union of the *costocervical* and *vertebral veins.*

External jugular vein. Formed by the union of the anterior and posterior facial veins. It returns blood from the head.

Internal jugular vein. A tributary of the external jugular vein which returns blood from the brain and spinal cord.

Internal thoracic veins. Paired veins which lie on either side of the midline on the inner surface of the ventral thoracic wall. They join to enter the ventral aspect of the precava by a common stem.

Precava. Returns unoxygenated blood from the head, the forelimbs, and the cranial part of the thorax to the right atrium. It receives the azygos and internal thoracic veins.

Subclavian vein. A tributary of the brachiocephalic vein. It is continuous with the *axillary vein* and receives the *subscapular vein* from the shoulder muscles.

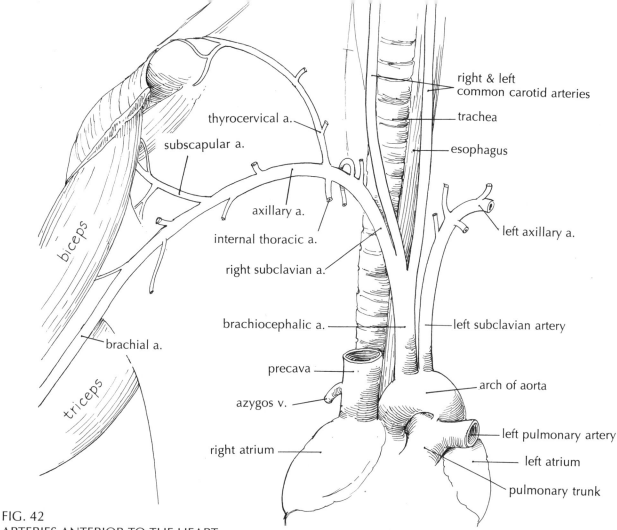

FIG. 42
ARTERIES ANTERIOR TO THE HEART

Remove the precava and its tributaries. Identify the arteries illustrated in Figure 42.

Aorta. The main arterial trunk of the body. It begins at the aortic valve, ascends for a short distance (*ascending aorta*), curves to the left, forming the *arch of the aorta,* and then, as the *descending aorta,* passes caudally. The descending aorta consists of *thoracic* and *abdominal* parts.

The ascending aorta gives off the *right* and *left coronary arteries.*

The arch of the aorta gives off the *brachiocephalic* and *left subclavian arteries.*

The thoracic aorta lies to the left of the vertebral column and dorsal to the heart. It extends from the arch of the aorta to the aortic hiatus of the diaphragm, giving off branches to the pericardium, bronchi, and esophagus. It also gives off paired *intercostal arteries,* each of which passes to an intercostal space.

Axillary artery. The continuation of the subclavian artery in the axilla. It gives off branches to the muscles of the shoulder and thorax, and then continues as the *brachial artery* to supply the muscles of the upper forelimb. Distal to the elbow it divides into *radial* and *ulnar* arteries, which supply the lower forelimb.

Brachiocephalic artery. Arises from the arch of the aorta. At the level of the second rib it gives rise to the right and left common carotid arteries and then continues as the *right subclavian artery.*

Common carotid artery. Gives off branches to the neck muscles, thyroid gland, and larynx. It then gives off the *internal carotid artery,* which supplies the brain, and continues as the *external carotid artery* to supply the neck, tongue, and face.

Left subclavian artery. Arises from the arch of the aorta, curves around the first rib, and continues as the *axillary artery.*

Right subclavian artery. A branch of the brachiocephalic artery. It gives off branches to the neck and shoulder, and continues as the *axillary artery.*

51

Remove abdominal viscera, connective tissue, fat, and peritoneum to expose the abdominal aorta and postcava. Identify the structures illustrated in Figure 43.

Abdominal aorta. That portion of the aorta which extends from the aortic hiatus of the diaphragm to the pelvis. It lies to the left of the midline and ventral to the vertebral column. At the level of the sacrum it gives off the paired external iliac arteries and continues in the midline to supply the pelvic viscera.

Celiac artery. Arises from the ventral aspect of the aorta just posterior to the aortic hiatus of the diaphragm and gives off three branches: hepatic, left gastric, and splenic.

External iliac arteries. Paired terminal branches of the abdominal aorta. They give off branches to the pelvis and hip, continuing into the leg as the femoral arteries.

Hepatic artery. The branch of the celiac artery which supplies the liver and gives branches to the stomach, pancreas, and duodenum.

Hepatic portal system of veins. The veins which convey blood from the abdominal viscera, via *the hepatic portal vein,* to capillary-like vessels (termed *hepatic sinusoids*) within the liver.

Hepatic veins. Several variable veins which return blood from the liver to the postcava.

Iliolumbar arteries. Paired arteries which arise from the aorta and pass laterally on either side to supply the muscles of the lumbar region.

Inferior mesenteric artery. Supplies the descending colon and rectum.

Left gastric artery. The branch of the celiac artery which supplies the lesser curvature of the stomach.

Lumbar arteries. Seven paired arteries which arise from the dorsal aspect of the aorta. They supply the muscles of the dorsal body wall and the vertebral column.

Ovarian arteries. Paired arteries which arise from the aorta and pass laterally in the broad ligament to supply the ovaries and the uterus.

Phrenicoabdominal arteries. Paired arteries which arise from the aorta just posterior to the superior mesenteric artery. They supply the muscles of the dorsal body wall and give branches to the diaphragm and the suprarenal glands.

Postcava. Returns blood from the body posterior to the diaphragm. It is formed at the level of the sacrum by the union of the right and left common iliac veins. It lies to the right of the aorta and passes through the caudate lobe of the liver, within which it receives the hepatic veins. It then passes through the postcaval foramen of the diaphragm and opens into the right atrium.

Renal arteries. Paired arteries which arise from the aorta and supply the kidneys.

Splenic artery. The branch of the celiac artery which supplies the spleen and gives branches to the pancreas and greater omentum.

Superior mesenteric artery. Supplies the small intestine, pancreas, duodenum, and ascending and transverse parts of the colon.

Testicular arteries. Paired arteries which arise from the aorta and pass posteriorly to supply the testicles.

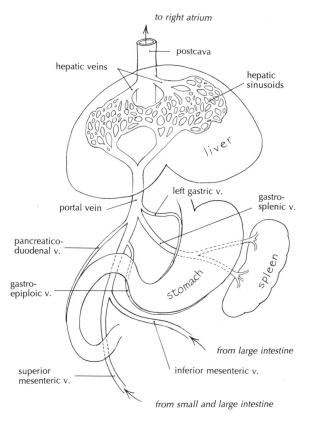

The Hepatic Portal System

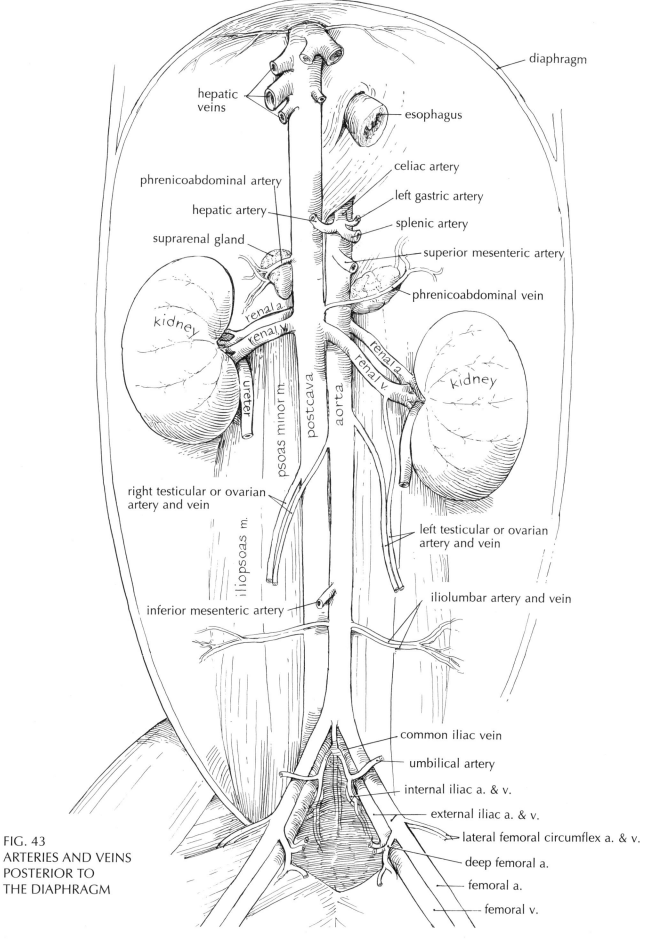

hepatic veins

diaphragm

esophagus

celiac artery

left gastric artery

splenic artery

phrenicoabdominal artery

hepatic artery

superior mesenteric artery

suprarenal gland

phrenicoabdominal vein

kidney

renal a.

renal v.

renal a.

renal v.

ureter

psoas minor m.

postcava

aorta

kidney

right testicular or ovarian artery and vein

iliopsoas m.

left testicular or ovarian artery and vein

inferior mesenteric artery

iliolumbar artery and vein

common iliac vein

umbilical artery

internal iliac a. & v.

external iliac a. & v.

lateral femoral circumflex a. & v.

deep femoral a.

femoral a.

femoral v.

FIG. 43
ARTERIES AND VEINS
POSTERIOR TO
THE DIAPHRAGM

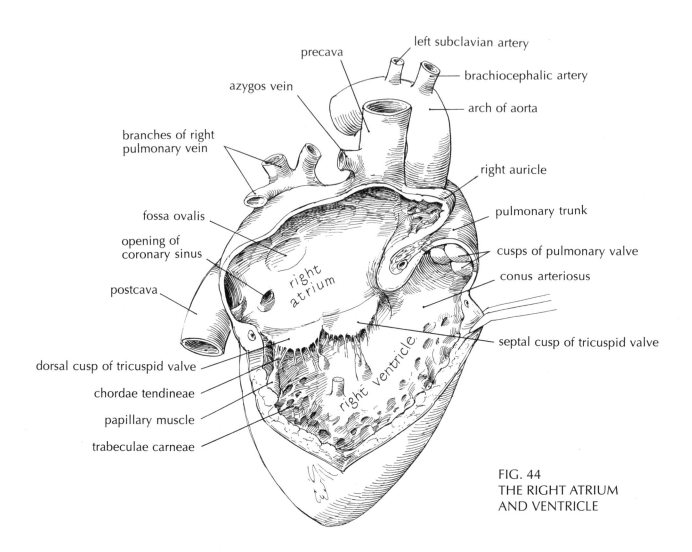

precava

azygos vein

branches of right
pulmonary vein

fossa ovalis

opening of
coronary sinus

postcava

dorsal cusp of tricuspid valve

chordae tendineae

papillary muscle

trabeculae carneae

left subclavian artery

brachiocephalic artery

arch of aorta

right auricle

pulmonary trunk

cusps of pulmonary valve

conus arteriosus

septal cusp of tricuspid valve

right atrium

right ventricle

FIG. 44
THE RIGHT ATRIUM
AND VENTRICLE

Remove the heart and trim away portions of the lateral wall of the right atrium. Pass a probe through the precava and observe that it enters the right atrium. Similarly, pass a probe through the postcava and observe that it, too, enters the right atrium. Trim away the lateral and ventral walls of the right ventricle. Pass a probe into the pulmonary trunk. Identify the structures illustrated in Figure 44.

Chordae tendineae. Fibrous cords which attach the free borders of the tricuspid and bicuspid valves to the interventricular septum and to muscular projections of the ventricular wall termed *papillary muscles.*

Foramen ovale. An opening in the interatrial septum of the fetal heart.

Fossa ovalis. The smooth oval depression in the medial wall of the right atrium. It is the site of the foramen ovale in the fetal heart.

Musculi pectinati. Ridges of muscle fibers on the inner walls of the auricles.

Opening of the coronary sinus. An opening in the dorsal wall of the right atrium. It is guarded by a semilunar valve, and conveys blood from the coronary sinus to the right atrium.

Pulmonary valve. Three pocket-like cusps which guard the origin of the pulmonary trunk.

Right atrioventricular opening. The opening which conveys unoxygenated blood from the right atrium to the right ventricle. It is guarded by the tricuspid valve.

Trabeculae carneae. Irregular muscular strands on the inner walls of the ventricles.

Tricuspid valve. Consists of three thin membranous cusps, termed *dorsal, ventral,* and *septal.*

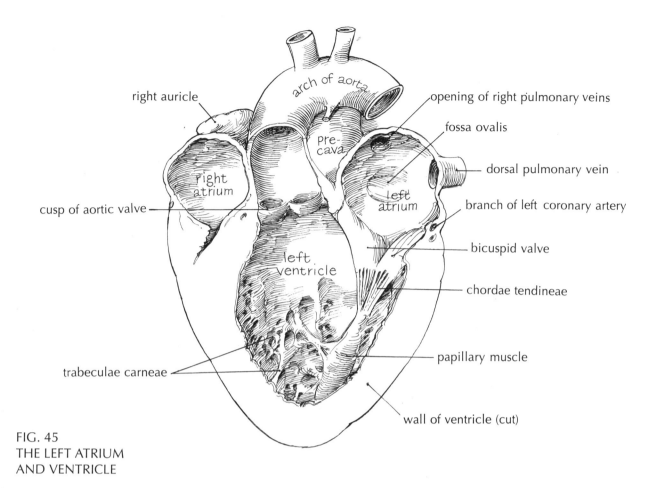

right auricle

arch of aorta

opening of right pulmonary veins

fossa ovalis

pre-cava

right atrium

dorsal pulmonary vein

left atrium

cusp of aortic valve

branch of left coronary artery

left ventricle

bicuspid valve

chordae tendineae

papillary muscle

trabeculae carneae

wall of ventricle (cut)

FIG. 45
THE LEFT ATRIUM
AND VENTRICLE

Remove the pulmonary trunk and trim away the lateral walls of the left atrium and ventricle. Trace the pulmonary veins to their openings in the dorsal wall of the left atrium. Cut the medial wall of the left atrium and make a slit in the lateral wall of the aorta, observing the three cusps of the aortic valve. Identify the structures illustrated in Figure 45.

Left atrioventricular opening. The opening which conveys oxygenated blood from the left atrium to the left ventricle. It is guarded by the bicuspid valve.

Bicuspid (mitral) valve. Consists of two thin membranous cusps (termed *septal* and *lateral*).

Aortic valve. Three pocket-like cusps at the origin of the aorta.

Interatrial septum. The muscular septum which separates the right and left atria.

Interventricular septum. The muscular septum which separates the right and left ventricles.

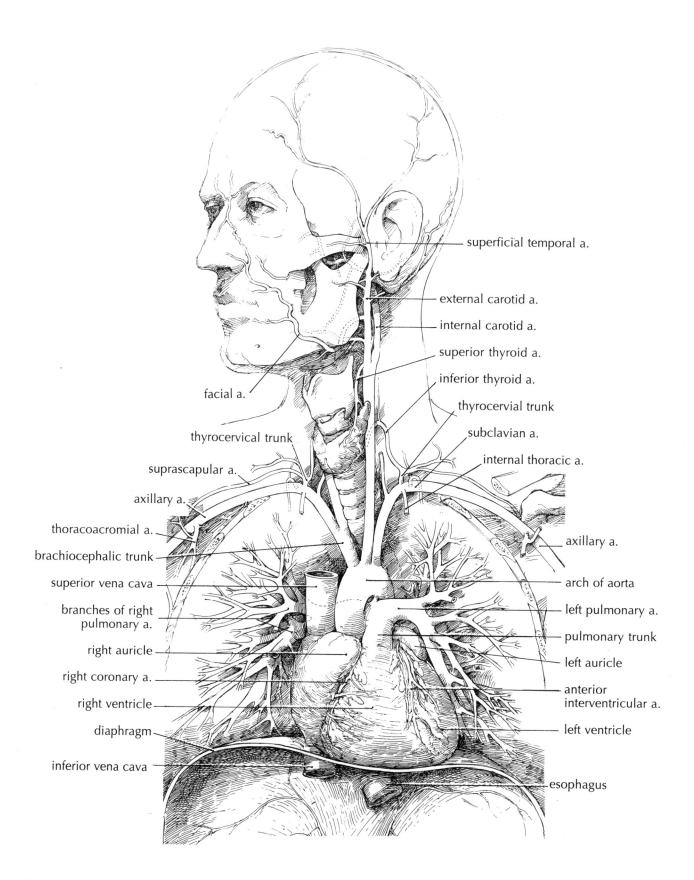

superficial temporal a.

external carotid a.

internal carotid a.

superior thyroid a.

inferior thyroid a.

thyrocervial trunk

subclavian a.

internal thoracic a.

facial a.

thyrocervical trunk

suprascapular a.

axillary a.

thoracoacromial a.

brachiocephalic trunk

superior vena cava

branches of right pulmonary a.

right auricle

right coronary a.

right ventricle

diaphragm

inferior vena cava

axillary a.

arch of aorta

left pulmonary a.

pulmonary trunk

left auricle

anterior interventricular a.

left ventricle

esophagus

FIG. 46
THE HUMAN HEART AND ARTERIES
OF THE HEAD AND SHOULDER

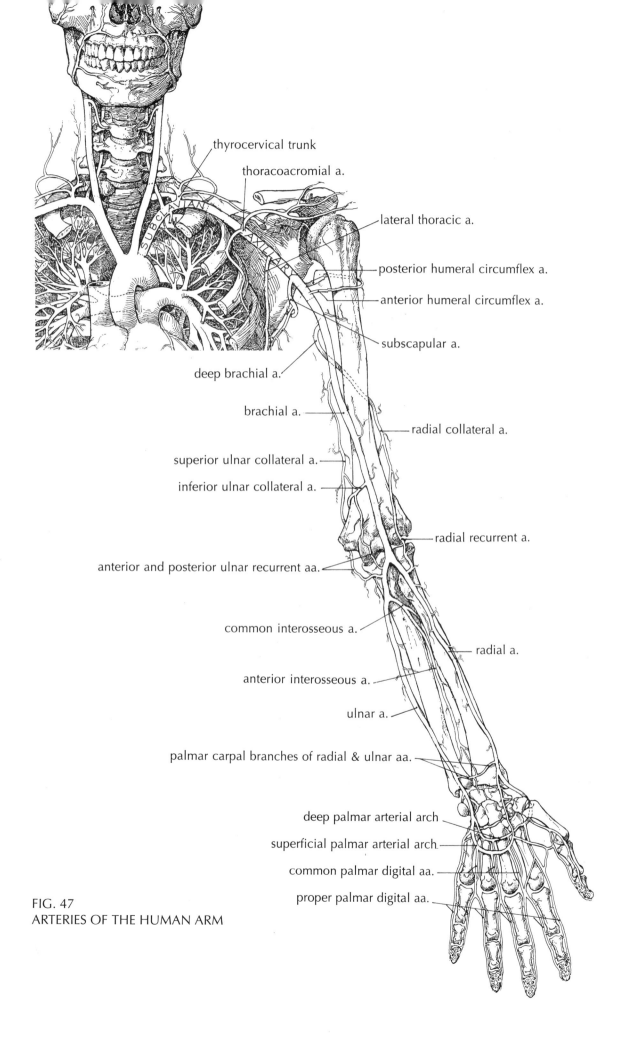

thyrocervical trunk

thoracoacromial a.

lateral thoracic a.

posterior humeral circumflex a.

anterior humeral circumflex a.

subscapular a.

deep brachial a.

brachial a.

radial collateral a.

superior ulnar collateral a.

inferior ulnar collateral a.

radial recurrent a.

anterior and posterior ulnar recurrent aa.

common interosseous a.

radial a.

anterior interosseous a.

ulnar a.

palmar carpal branches of radial & ulnar aa.

deep palmar arterial arch

superficial palmar arterial arch

common palmar digital aa.

proper palmar digital aa.

FIG. 47
ARTERIES OF THE HUMAN ARM

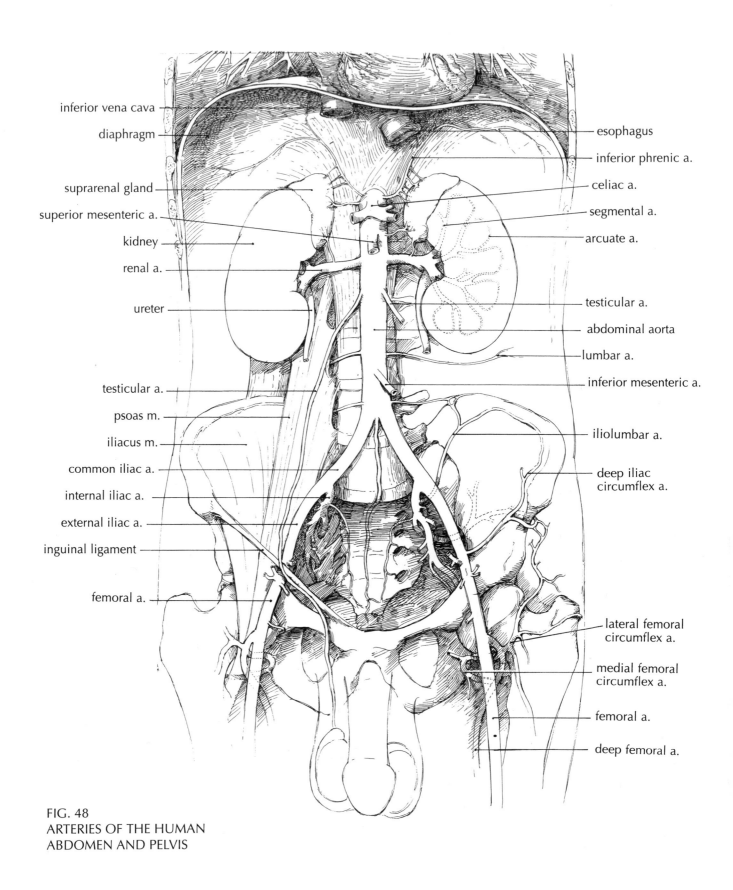

inferior vena cava

diaphragm

suprarenal gland

superior mesenteric a.

kidney

renal a.

ureter

testicular a.

psoas m.

iliacus m.

common iliac a.

internal iliac a.

external iliac a.

inguinal ligament

femoral a.

esophagus

inferior phrenic a.

celiac a.

segmental a.

arcuate a.

testicular a.

abdominal aorta

lumbar a.

inferior mesenteric a.

iliolumbar a.

deep iliac
circumflex a.

lateral femoral
circumflex a.

medial femoral
circumflex a.

femoral a.

deep femoral a.

FIG. 48
ARTERIES OF THE HUMAN
ABDOMEN AND PELVIS

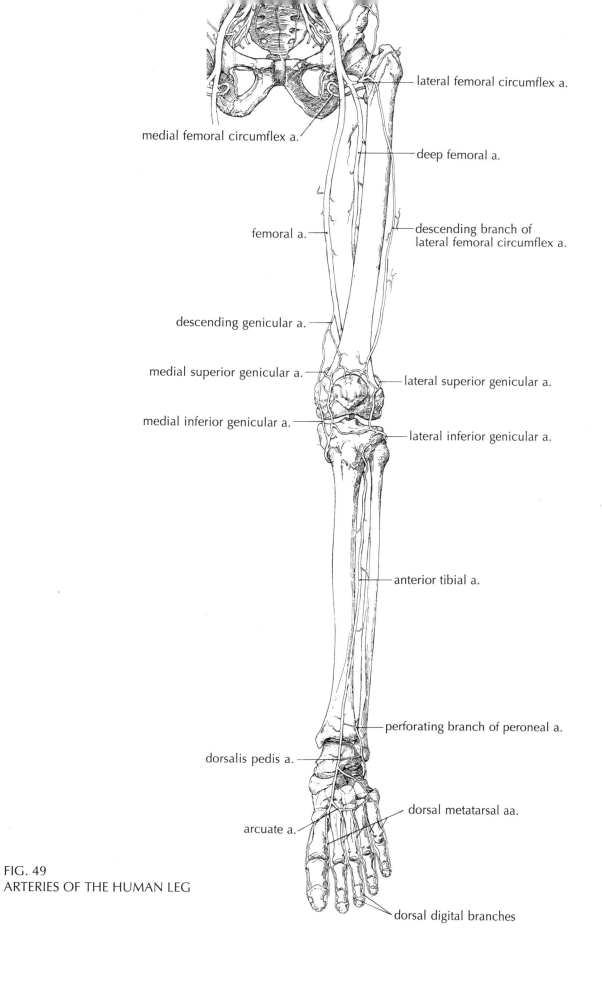

lateral femoral circumflex a.

medial femoral circumflex a.

deep femoral a.

femoral a.

descending branch of
lateral femoral circumflex a.

descending genicular a.

medial superior genicular a.

lateral superior genicular a.

medial inferior genicular a.

lateral inferior genicular a.

anterior tibial a.

perforating branch of peroneal a.

dorsalis pedis a.

dorsal metatarsal aa.

arcuate a.

FIG. 49
ARTERIES OF THE HUMAN LEG

dorsal digital branches

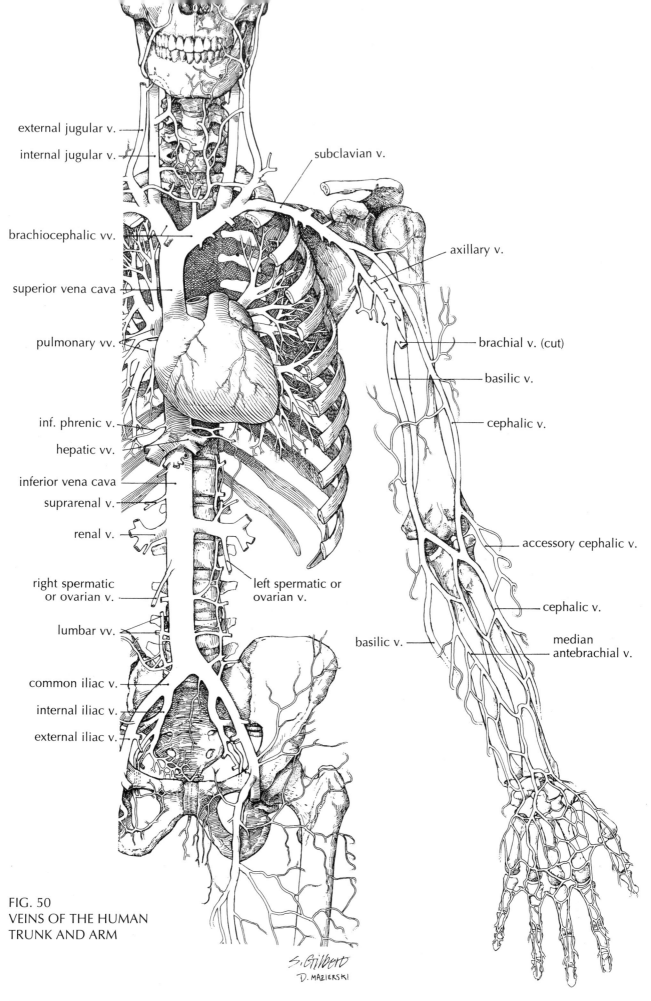

external jugular v.

internal jugular v.

subclavian v.

brachiocephalic vv.

axillary v.

superior vena cava

pulmonary vv.

brachial v. (cut)

basilic v.

cephalic v.

inf. phrenic v.

hepatic vv.

inferior vena cava

suprarenal v.

renal v.

accessory cephalic v.

right spermatic
or ovarian v.

left spermatic or
ovarian v.

cephalic v.

lumbar vv.

median
antebrachial v.

basilic v.

common iliac v.

internal iliac v.

external iliac v.

FIG. 50
VEINS OF THE HUMAN
TRUNK AND ARM

S. Gilbert
D. MAZIERSKI

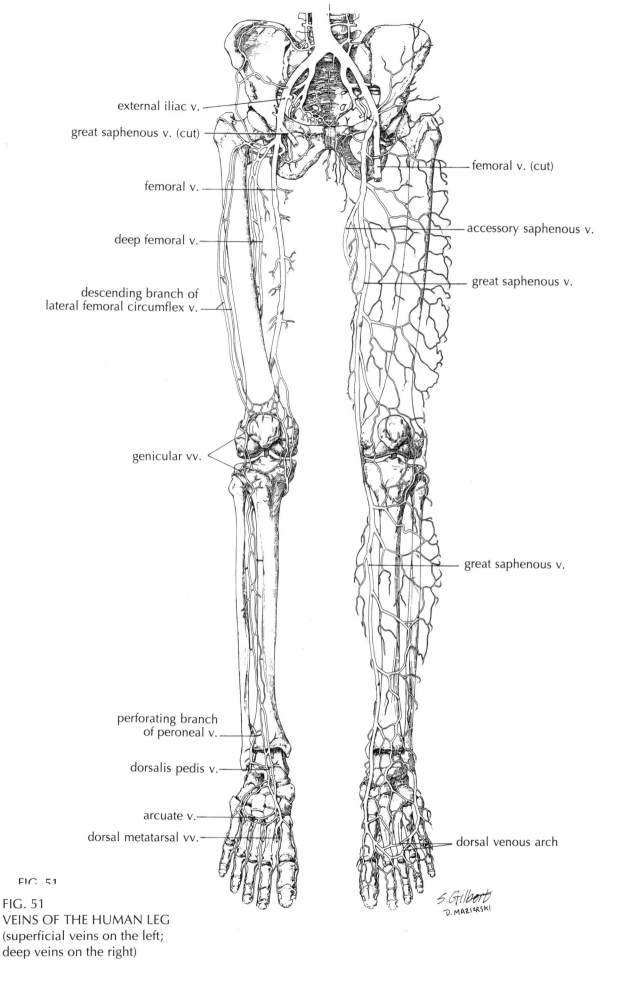

external iliac v.

great saphenous v. (cut)

femoral v.

deep femoral v.

descending branch of
lateral femoral circumflex v.

genicular vv.

perforating branch
of peroneal v.

dorsalis pedis v.

arcuate v.

dorsal metatarsal vv.

femoral v. (cut)

accessory saphenous v.

great saphenous v.

great saphenous v.

dorsal venous arch

S. Gilbert
D. MAZIERSKI

FIG. 51

FIG. 51
VEINS OF THE HUMAN LEG
(superficial veins on the left;
deep veins on the right)

THE NERVOUS SYSTEM

Central Nervous System
> brain
> spinal cord

Peripheral Nervous System
> 12 pairs of cranial nerves
> 38 pairs of spinal nerves

Autonomic Nervous System
> sympathetic system
> parasympathetic system

SPINAL CORD

Remove the muscles dorsal to the vertebral column in the upper lumbar region. Use small bone clippers to remove the neural arches of several vertebrae and expose the spinal cord together with the roots of several spinal nerves.

Central canal. The minute passage which extends from the fourth ventricle throughout the length of the spinal cord.

Cerebrospinal fluid. A clear fluid similar to lymph. It is elaborated by the *choroid plexus* of the cerebral ventricles.

Cervical enlargement of the spinal cord. An enlargement of the spinal cord which extends from C-4 to T-1. The nerves which innervate the forelimb originate from the cervical enlargement.

Conus medullaris. The conical caudal end of the spinal cord.

Filum terminale. The fibrous filament which extends from the apex of the conus medullaris into the vertebral canal of the caudal vertebrae.

Gray matter. Nerve cells and closely related processes.

Lumbar enlargement of the spinal cord. An enlargement of the spinal cord which extends from L-3 to L-7. The nerves which innervate the hindlimb originate from this part of the lumbar enlargement.

Meninges. Three membranes which enclose the brain and spinal cord. They are termed the *dura mater, the arachnoid,* and the *pia mater.*

> **Dura mater**. The strongest and most superficial of the three membranes. It consists of a dense fibrous connective tissue.

> **Arachnoid**. A web-like membrane which, together with the cerebrospinal fluid, occupies the space between the dura mater and the pia mater.

> **Pia mater**. The thin innermost membrane. It carries a network of delicate blood vessels and is closely applied to surface of the brain and spinal cord.

Myelinated nerve fibers. Nerve fibers which are enclosed in a sheath composed of a fat-like substance termed *myelin.*

Spinal cord. That part of the central nervous system which lies within the vertebral canal. Cranially, it is continuous with the medulla oblongata; caudally, it ends in the filum terminale.

Spinal nerve. Thirty-one pairs of spinal nerves arise from the spinal cord. In the cat they include 8 pairs of cervical nerves, 13 pairs of thoracic nerves, 7 pairs of lumbar nerves, and 3 pairs of sacral nerves.

White matter. Bundles or masses of nerve fibers, mostly myelinated.

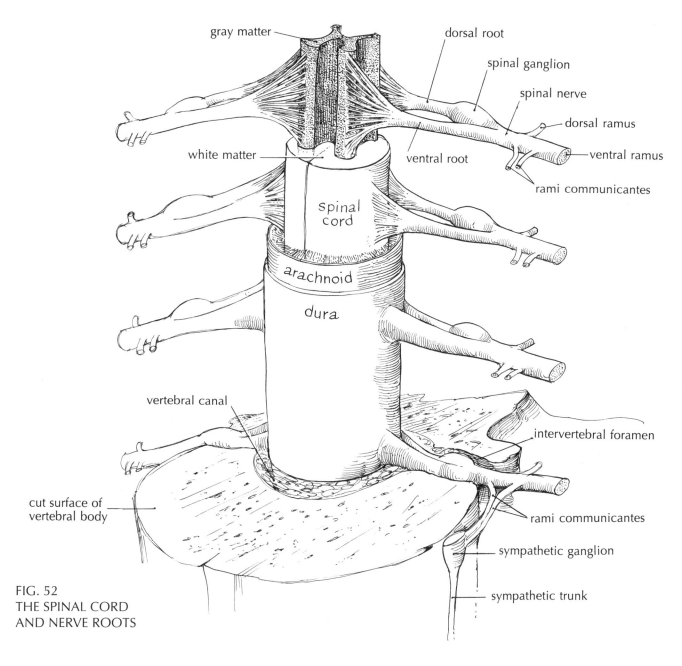

FIG. 52
THE SPINAL CORD
AND NERVE ROOTS

SPINAL NERVES

Each *spinal nerve* arises from the spinal cord by *dorsal* and *ventral roots.*

Dorsal root. The sensory division of each spinal nerve. Each dorsal root bears a **spinal ganglion** which contains cell bodies of sensory nerve fibers.

Ventral root. The motor division of each spinal nerve. Cell bodies of motor nerve fibers lie within the spinal cord. Dorsal and ventral roots join to form the spinal nerve, which emerges through the intervertebral foramen.

Dorsal ramus. The branch of a spinal nerve which supplies the skin and muscles on the dorsal aspect of the neck and trunk. It is considerably smaller than the ventral ramus.

Ventral ramus. The larger branch of a spinal nerve which is distributed to the limbs and to the skin and muscles on the ventral and lateral parts of the trunk. Ventral rami contribute to cervical, brachial, and lumbar plexuses.

Rami communicantes. Sympathetic nerve fibers which connect sympathetic ganglia and spinal nerves.

63

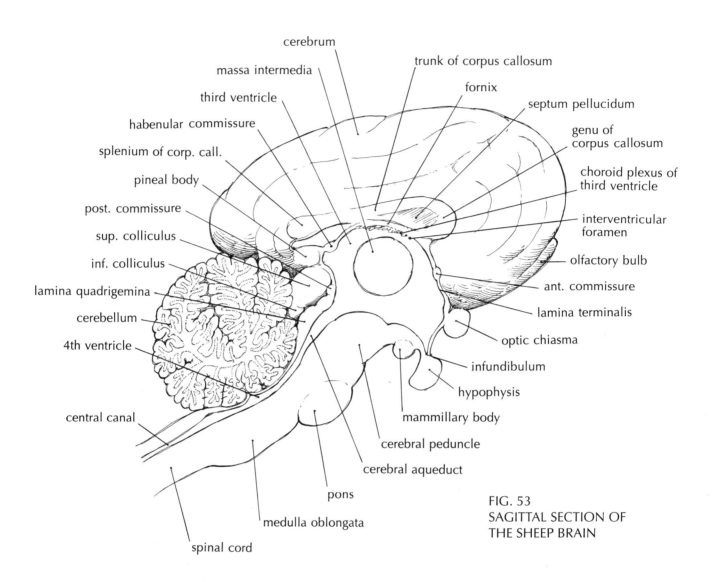

cerebrum
massa intermedia
third ventricle
habenular commissure
splenium of corp. call.
pineal body
post. commissure
sup. colliculus
inf. colliculus
lamina quadrigemina
cerebellum
4th ventricle
central canal

trunk of corpus callosum
fornix
septum pellucidum
genu of corpus callosum
choroid plexus of third ventricle
interventricular foramen
olfactory bulb
ant. commissure
lamina terminalis
optic chiasma
infundibulum
hypophysis
mammillary body
cerebral peduncle
cerebral aqueduct
pons
medulla oblongata
spinal cord

FIG. 53
SAGITTAL SECTION OF
THE SHEEP BRAIN

TELENCEPHALON OR CEREBRUM. The largest part of the brain. It consists of *right* and *left cerebral hemispheres* united by a transverse band of fibers termed the *corpus callosum.*

> **Basal ganglia.** The central gray matter of the cerebrum.
>
> **Central white matter.** The internal portion of the cerebrum (except the basal ganglia).
>
> **Cerebral cortex.** The peripheral gray matter of the cerebrum.
>
> **Neopallium.** The non-olfactory part of the cerebral cortex.
>
> **Rhinencephalon.** Structures functional in the reception and conduction of olfactory sensations: *olfactory bulbs, tracts, striae,* and *piriform lobes.*

DIENCEPHALON. Structures which surround the third ventricle and serve to connect the cerebral hemispheres with the mesencephalon.

> **Epithalamus.** The *pineal body, posterior commissure,* and *habenula.*
>
> **Hypothalamus.** Structures which form most of the floor and lower part of the lateral walls of the third ventricle, Including the *mammillary bodies, tuber cinereum, infundibulum, hypophysis,* and *optic chiasma.*
>
> **Thalamus.** Bilaterally symmetrical ovoid masses of gray matter which lie on either side of the third ventricle, forming most of its lateral walls.

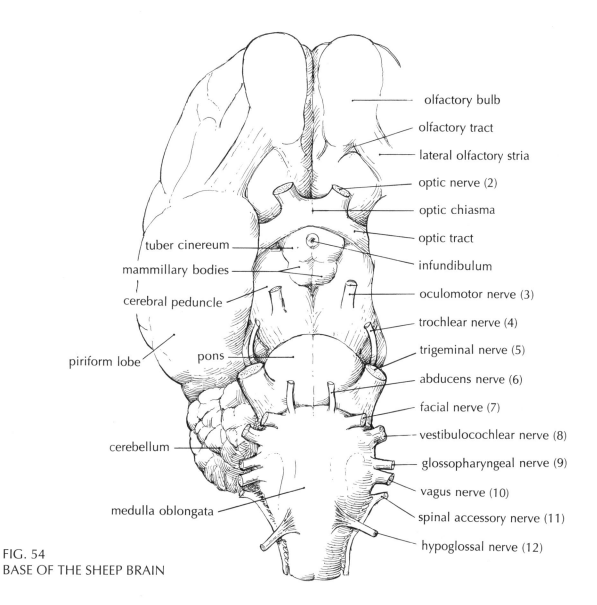

olfactory bulb
olfactory tract
lateral olfactory stria
optic nerve (2)
optic chiasma
optic tract
infundibulum
oculomotor nerve (3)
trochlear nerve (4)
trigeminal nerve (5)
abducens nerve (6)
facial nerve (7)
vestibulocochlear nerve (8)
glossopharyngeal nerve (9)
vagus nerve (10)
spinal accessory nerve (11)
hypoglossal nerve (12)

tuber cinereum
mammillary bodies
cerebral peduncle
piriform lobe
pons
cerebellum
medulla oblongata

FIG. 54
BASE OF THE SHEEP BRAIN

MESENCEPHALON. A short segment of the brain stem which connects the pons and cerebellum with the telencephalon.

   Cerebral aqueduct. A channel in the mesencephalon which connects the third and fourth ventricles.

   Cerebral peduncles. Fiber tracts which connect the pons with the diencephalon and telencephalon.

   Corpora quadrigemina. Four rounded prominences, consisting of superior and inferior colliculi, on the superior aspect of the mesencephalon.

METENCEPHALON. Consists of the pons and the cerebellum.

   Arbor vitae. The tree-like pattern of gray and white matter seen in section of the cerebellum.

Cerebellar peduncles. Fiber tracts which connect the cerebellum to the mesencephalon.

Cerebellum. The division of the brain which lies dorsal to the fourth ventricle and posterior to the cerebrum. It consists of central white matter covered by a convoluted cortex of gray matter.

Pons. A prominent swelling on the ventral aspect of the brain stem between the medulla oblongata and the cerebral peduncles.

MYELECEPHALON OR MEDULLA OBLONGATA. That portion of the brain stem which lies between the pons and the first spinal nerve. It forms the floor of the fourth ventricle.

65

Nerve	Superficial origin	Foramen	Distribution	Function
1. **Olfactory**	olfactory bulb	foramina in cribiform plate of ethmoid bone	olfactory mucosa	smell
2. **Optic**	optic chiasma	optic foramen	retina	vision
3. **Oculomotor**	cerebral peduncle	orbital fissure	levator palpebrae, sup. rectus, med.. rectus, inf. rectus, inf. Oblique	ocular movement
4. **Trochlear**	roof of 4th ventricle (ant. medullary velum)	orbital fissure	superior oblique	ocular movement
5. **Trigeminal**	pons			
ophthalmic		orbital fissure	eyelid, nasal mucosa	sensory
maxillary		foramen rotundum	mouth and face	sensory
mandibular		foramen ovale	muscles of mastication; mouth and face	jaw movement, facial sensation
6. **Abducens**	between pons and medulla	orbital fissure	lateral rectus, retractor oculi	ocular movement
7. **Facial**	pons	internal auditory meatus; facial canal; stylomastoid foramen	facial muscles; stapedius, stylohyoid, digastric, tongue,	movement of facial muscles, glandular secretion, taste
8. **Vestibulo-cochlear**	medulla	internal auditory meatus		
vestibular			saccule, utricle, semicircular ducts	equilibrium
cochlear			cochlear duct	hearing
9. **Glosso-pharyngeal**	medulla	jugular foramen	stylopharyngeus; muscles and mucosa of pharynx	taste, swallowing, glandular secretion
10. **Vagus**	medulla	jugular foramen	pharynx; thoracic and abdominal viscera	swallowing, phonation, taste, involuntary muscles of viscera;
11. **Accessory**	medulla and spinal cord	jugular foramen	cleidomastoid, sterno-mastoid, trapezius; muscles of pharynx and larynx	movement of shoulder and head; swallowing and phonation
12. **Hypoglossal**	medulla	hypoglossal canal	hyoid and tongue muscles	movements of tongue

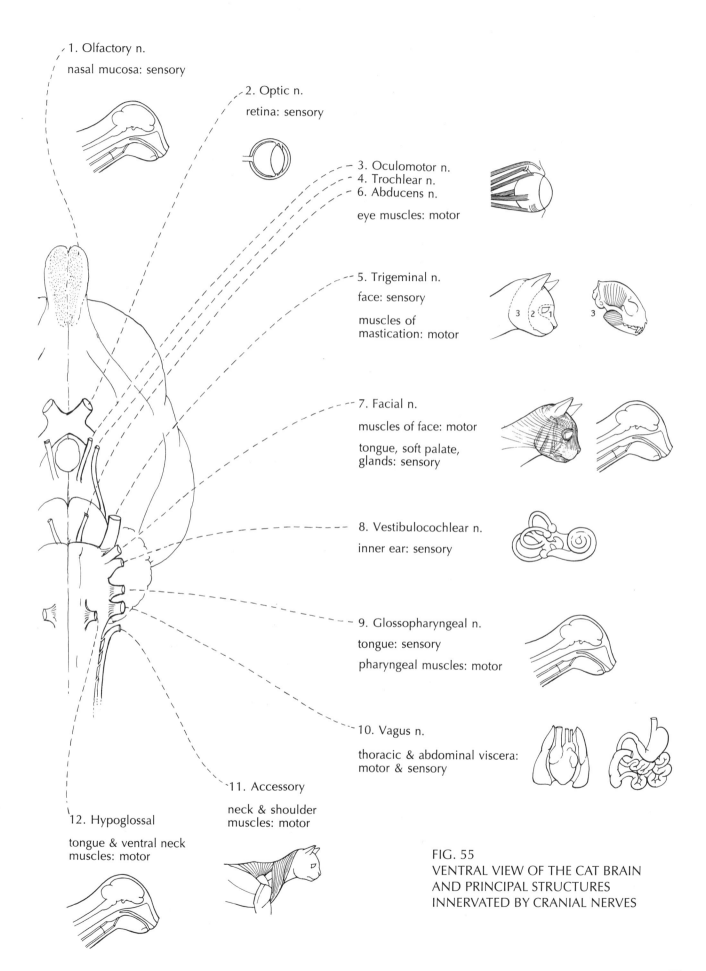

1. Olfactory n.

nasal mucosa: sensory

2. Optic n.

retina: sensory

3. Oculomotor n.
4. Trochlear n.
6. Abducens n.

eye muscles: motor

5. Trigeminal n.

face: sensory

muscles of
mastication: motor

7. Facial n.

muscles of face: motor

tongue, soft palate,
glands: sensory

8. Vestibulocochlear n.

inner ear: sensory

9. Glossopharyngeal n.

tongue: sensory

pharyngeal muscles: motor

10. Vagus n.

thoracic & abdominal viscera:
motor & sensory

11. Accessory

neck & shoulder
muscles: motor

12. Hypoglossal

tongue & ventral neck
muscles: motor

FIG. 55
VENTRAL VIEW OF THE CAT BRAIN
AND PRINCIPAL STRUCTURES
INNERVATED BY CRANIAL NERVES

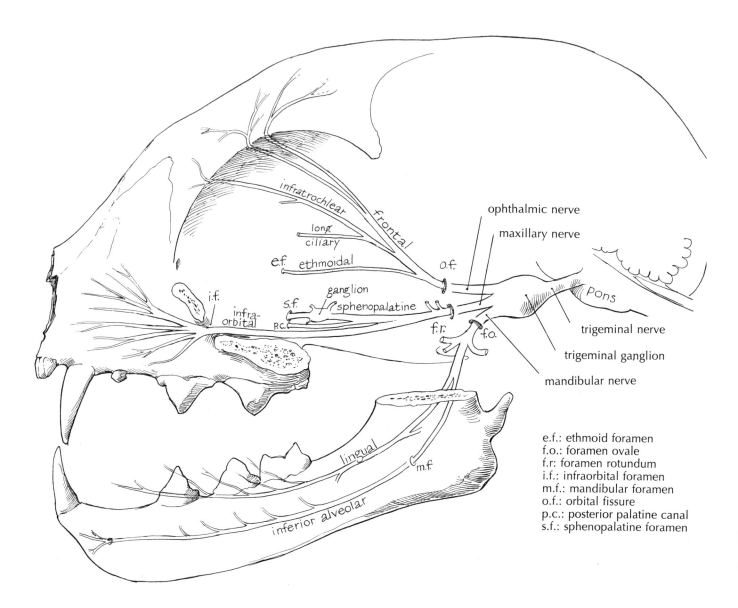

Labels on figure:
infratrochlear
frontal
long ciliary
e.f. ethmoidal
ganglion sphenopalatine
s.f.
p.c.
i.f.
infra-orbital
o.f.
f.r.
f.o.
ophthalmic nerve
maxillary nerve
Pons
trigeminal nerve
trigeminal ganglion
mandibular nerve
lingual
m.f.
inferior alveolar

e.f.: ethmoid foramen
f.o.: foramen ovale
f.r: foramen rotundum
i.f.: infraorbital foramen
m.f.: mandibular foramen
o.f.: orbital fissure
p.c.: posterior palatine canal
s.f.: sphenopalatine foramen

FIG. 56
THE TRIGEMINAL NERVE

Refer to a skull and identify the foramina illustrated in Figure 56. See a demonstration dissection in which the branches of the trigeminal nerve are exposed.

Refer to Figure 57. On either side of the trachea identify the vagus nerve, the common carotid artery, the phrenic nerve, and the sympathetic trunk (see Fig. 60). Being careful not to injure the brachial plexus, the phrenic nerve, or the sympathetic nerve, follow the vagus nerve caudally through the neck and thorax, identifying the structures illustrated in Figure 57. Find the point where the vagus nerve passes through the diaphragm and follow it to the stomach.

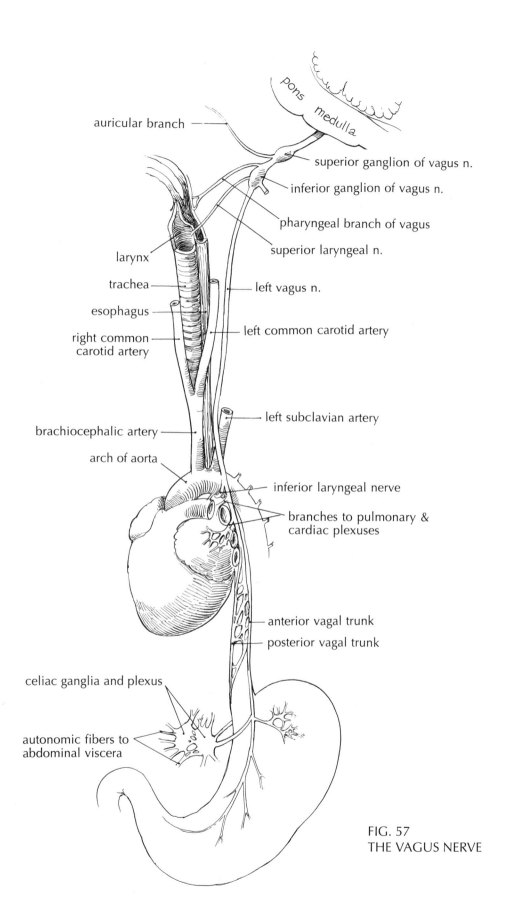

auricular branch

pons

medulla

superior ganglion of vagus n.

inferior ganglion of vagus n.

pharyngeal branch of vagus

superior laryngeal n.

larynx

trachea

esophagus

right common
carotid artery

left vagus n.

left common carotid artery

brachiocephalic artery

arch of aorta

left subclavian artery

inferior laryngeal nerve

branches to pulmonary &
cardiac plexuses

anterior vagal trunk

posterior vagal trunk

celiac ganglia and plexus

autonomic fibers to
abdominal viscera

FIG. 57
THE VAGUS NERVE

69

Nerve	Formation	Distribution
Ant. pectoral	C-7	pectoralis
Post. pectoral	C-8, T-1	pectoralis; sometimes latissimus dorsi
Long thoracic	C-7	serratus anterior
Suprascapular	C-6	supraspinaturs and infraspinatus
Subscapular #1	C-6, 7	subscapularis
Subscapular #2	C-7	teres major
Subscapular #3	C-7, 8	latissimus dorsi
Axillary	C-6-7	deltoids, clavobrachialis, teres minor
Phrenic	C-5, 6	diaphragm
Musculocutaneous	C-6-7	biceps, coracobrachialis, brachialis
Radial	C-7, 8; T-1	epitrochlearis, anconeus, triceps, supinator, and extensors of ulnar and dorsal side of forelimb
Median	C-7, 8; T-1	flexors of carpus and digits, except flexor carpi ulnaris, and ulnar head of flexor digitorum profundus
Ulnar	C-8; T-1	flexor carpi ulnaris, ulnar head of flexor digitorum profundus

Review and identify the muscles of the shoulder, chest, and upper forelimb. If the brachial plexus was damaged during the dissection of the muscles, work on the opposite side of the specimen. Remove muscles, veins, connective tissue, and fat to expose components of the brachial plexus as they emerge from the neck and thorax to enter the forelimb. Identify cervical nerves 5 through 8 and the first thoracic nerve. Follow the branches of the brachial plexus, identifying the nerves illustrated in Figure 58.

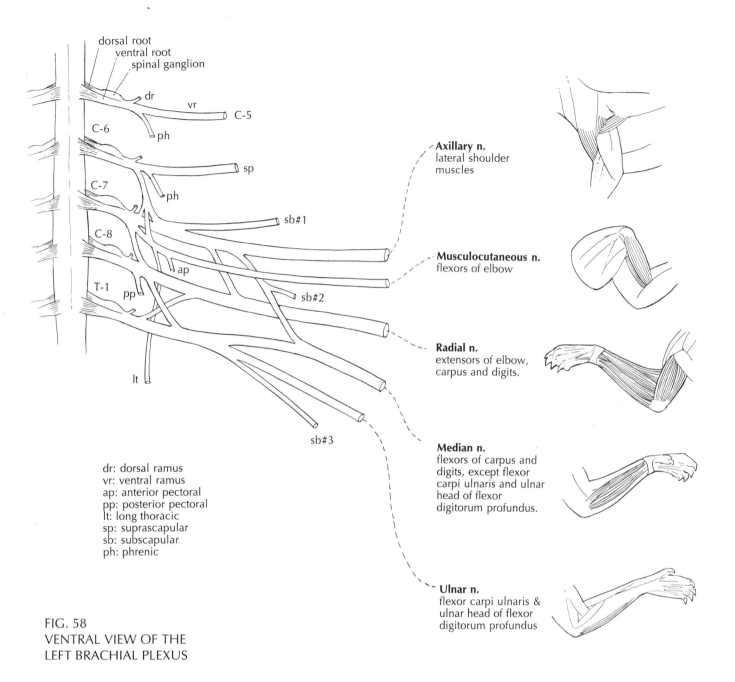

dorsal root
ventral root
spinal ganglion

dr
vr
C-5
ph

C-6
sp
ph

C-7
sb#1

C-8
ap

T-1
pp
sb#2

lt

sb#3

Axillary n.
lateral shoulder
muscles

Musculocutaneous n.
flexors of elbow

Radial n.
extensors of elbow,
carpus and digits.

Median n.
flexors of carpus and
digits, except flexor
carpi ulnaris and ulnar
head of flexor
digitorum profundus.

Ulnar n.
flexor carpi ulnaris &
ulnar head of flexor
digitorum profundus

dr: dorsal ramus
vr: ventral ramus
ap: anterior pectoral
pp: posterior pectoral
lt: long thoracic
sp: suprascapular
sb: subscapular
ph: phrenic

FIG. 58
VENTRAL VIEW OF THE
LEFT BRACHIAL PLEXUS

Nerve	Formation	Distribution
Genitofemoral	L-4	skin of thigh, genitals, and abdominal wall
Lateral femoral cutaneous	L-4, 5	skin of lateral surface of thigh
Femoral	L-5, 6	sartorius, quadriceps femoris
Obturator	L-6, 7	obturator externus, adductor femoris, adductor longus, pectineus and gracilis
Superior gluteal	lumbosacral cord (L-6, 7, S-1)	tensor fasciae latae, gluteus medius, gluteus minimus, gemellus superior
Inferior gluteal	lumbosacral cord and S-1	caudofemoralis and gluteus maximus
Sciatic	lumbosacral cord and S-1	biceps femoris, tenuissimus, semitendinosus, semimembranosus, quadratus femoris, branches to muscles of lower leg
Pudendal	S-2, 3	anal and genital region
Posterior femoral cutaneous	S-2, 3	perineum and skin of thigh
Inferior hemorrhoidal	S-2, 3	bladder and urethra

Remove the abdominal vessels and any remaining components of the urogenital system to expose the lumbar vertebrae and other structures of the dorsal abdominal wall. Identify the iliopsoas and psoas minor muscles. Cut and remove the psoas minor and the cranial part of the iliopsoas. Identify lumbar nerves 1, 2, and 3, which can be seen emerging from the intervertebral foramina between the corresponding lumbar vertebrae. Referring to Figure 59, identify the lateral femoral cutaneous nerve, the femoral nerve, and the obturator nerve, which lie on the ventral surface of the iliopsoas.

Examine the dorsal aspect of the thigh. Cut and retract the biceps femoris. The sciatic nerve, which is the largest nerve of the body, can be seen lying deep to the biceps femoris.

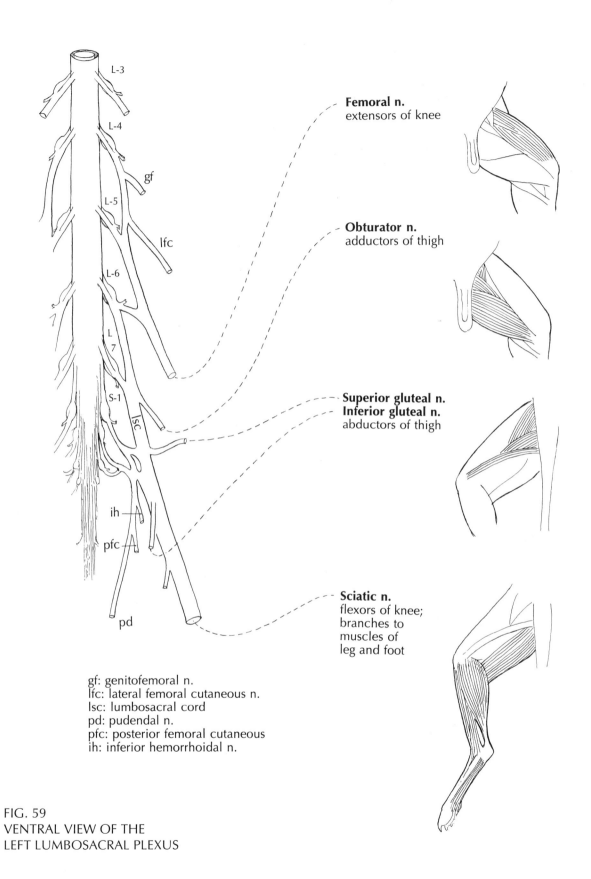

Femoral n.
extensors of knee

Obturator n.
adductors of thigh

Superior gluteal n.
Inferior gluteal n.
abductors of thigh

Sciatic n.
flexors of knee;
branches to
muscles of
leg and foot

gf: genitofemoral n.
lfc: lateral femoral cutaneous n.
lsc: lumbosacral cord
pd: pudendal n.
pfc: posterior femoral cutaneous
ih: inferior hemorrhoidal n.

FIG. 59
VENTRAL VIEW OF THE
LEFT LUMBOSACRAL PLEXUS

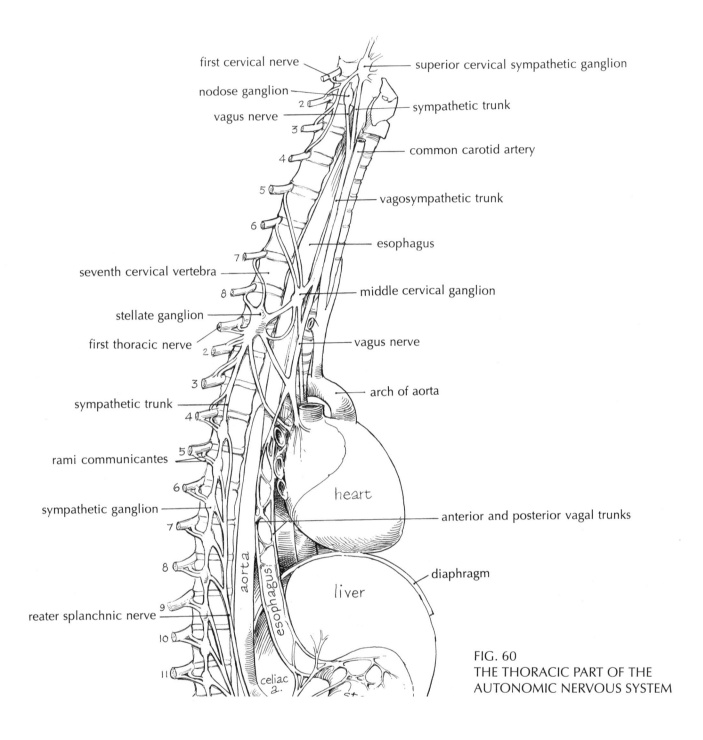

first cervical nerve

nodose ganglion

vagus nerve

seventh cervical vertebra

stellate ganglion

first thoracic nerve

sympathetic trunk

rami communicantes

sympathetic ganglion

reater splanchnic nerve

superior cervical sympathetic ganglion

sympathetic trunk

common carotid artery

vagosympathetic trunk

esophagus

middle cervical ganglion

vagus nerve

arch of aorta

anterior and posterior vagal trunks

diaphragm

heart

liver

aorta

esophagus

celiac a.

FIG. 60
THE THORACIC PART OF THE
AUTONOMIC NERVOUS SYSTEM

Autonomic nervous system. The system of ganglia, nerves, and plexuses which supplies the viscera, glands, blood vessels, and nonstriated muscles. It includes two complementary parts, termed *sympathetic* and *parasympathetic*, which are physiologically antagonistic to each other.

Sympathetic system. That part of the autonomic system which is connected with the central nervous system via thoracic and upper lumbar segments of the spinal cord. Sympathetic ganglia are situated near vertebral bodies.

Parasympathetic system. That part of the autonomic system which is connected with the central nervous system via certain cranial nerves and via sacral segments of the spinal cord. Parasympathetic ganglia are located near the organs innervated.

Sympathetic trunks. A series of ganglia connected by nerve cords which lie on either side of the vertebral bodies from the base of the skull to the sacrum.

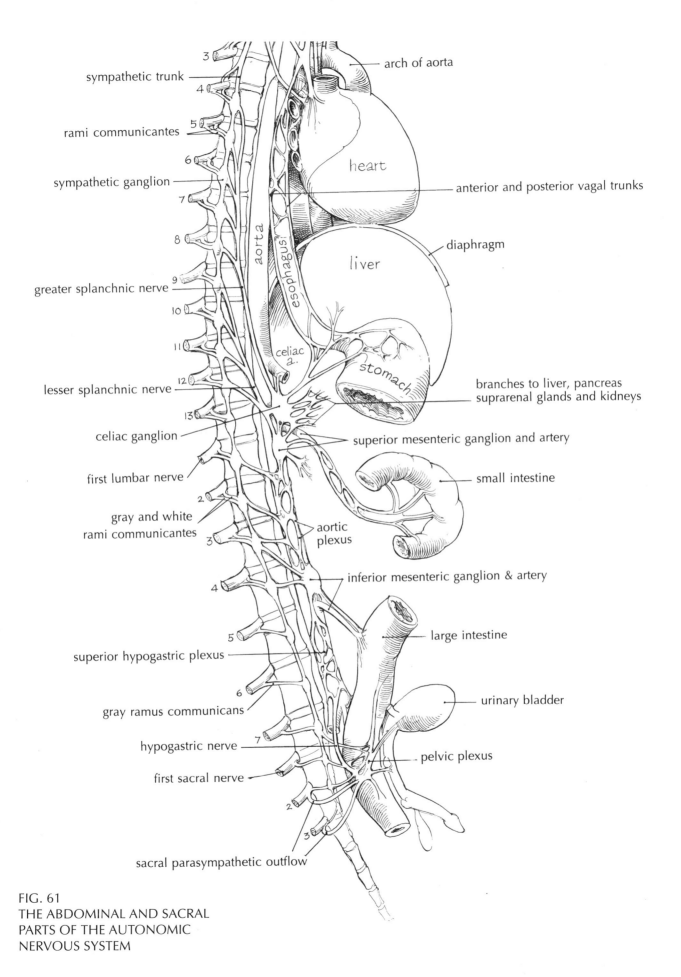

sympathetic trunk

rami communicantes

sympathetic ganglion

greater splanchnic nerve

lesser splanchnic nerve

celiac ganglion

first lumbar nerve

gray and white
rami communicantes

superior hypogastric plexus

gray ramus communicans

hypogastric nerve

first sacral nerve

sacral parasympathetic outflow

3
4
5
6
7
8
9
10
11
12
13

aorta

esophagus

celiac a.

2
3
4
5
6
7

2
3

arch of aorta

heart

liver

stomach

anterior and posterior vagal trunks

diaphragm

branches to liver, pancreas
suprarenal glands and kidneys

superior mesenteric ganglion and artery

small intestine

aortic
plexus

inferior mesenteric ganglion & artery

large intestine

urinary bladder

pelvic plexus

FIG. 61
THE ABDOMINAL AND SACRAL
PARTS OF THE AUTONOMIC
NERVOUS SYSTEM

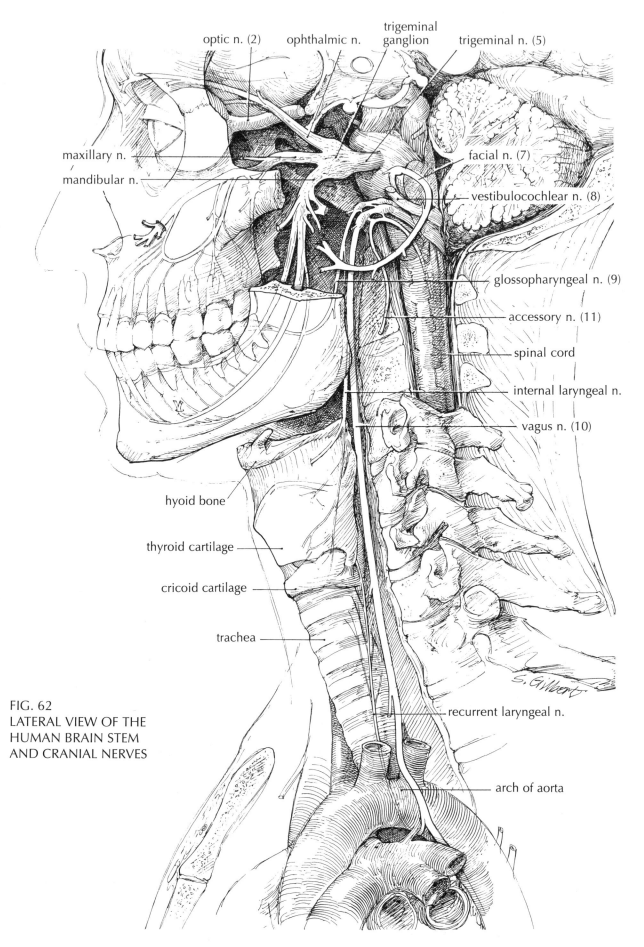

optic n. (2) ophthalmic n. trigeminal ganglion trigeminal n. (5)

maxillary n.

mandibular n.

facial n. (7)

vestibulocochlear n. (8)

glossopharyngeal n. (9)

accessory n. (11)

spinal cord

internal laryngeal n.

vagus n. (10)

hyoid bone

thyroid cartilage

cricoid cartilage

trachea

S. Gilbert

recurrent laryngeal n.

arch of aorta

FIG. 62
LATERAL VIEW OF THE
HUMAN BRAIN STEM
AND CRANIAL NERVES

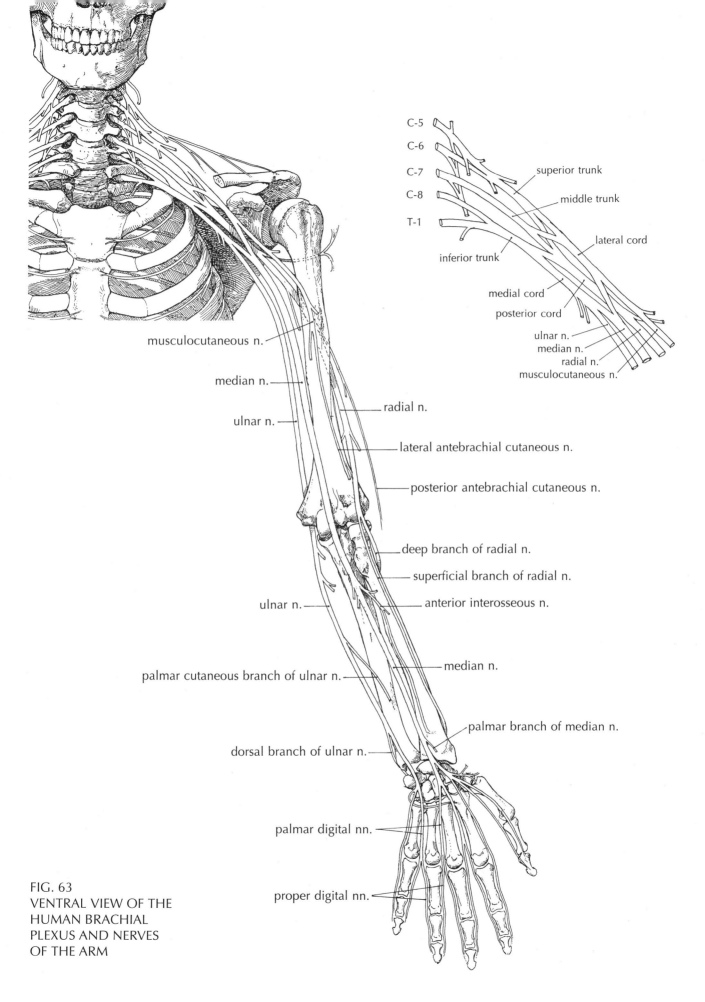

C-5
C-6
C-7
C-8
T-1

superior trunk

middle trunk

lateral cord

inferior trunk

medial cord
posterior cord

ulnar n.
median n.
radial n.
musculocutaneous n.

musculocutaneous n.

median n.

ulnar n.

radial n.

lateral antebrachial cutaneous n.

posterior antebrachial cutaneous n.

deep branch of radial n.

superficial branch of radial n.

anterior interosseous n.

ulnar n.

median n.

palmar cutaneous branch of ulnar n.

palmar branch of median n.

dorsal branch of ulnar n.

palmar digital nn.

proper digital nn.

FIG. 63
VENTRAL VIEW OF THE
HUMAN BRACHIAL
PLEXUS AND NERVES
OF THE ARM

77

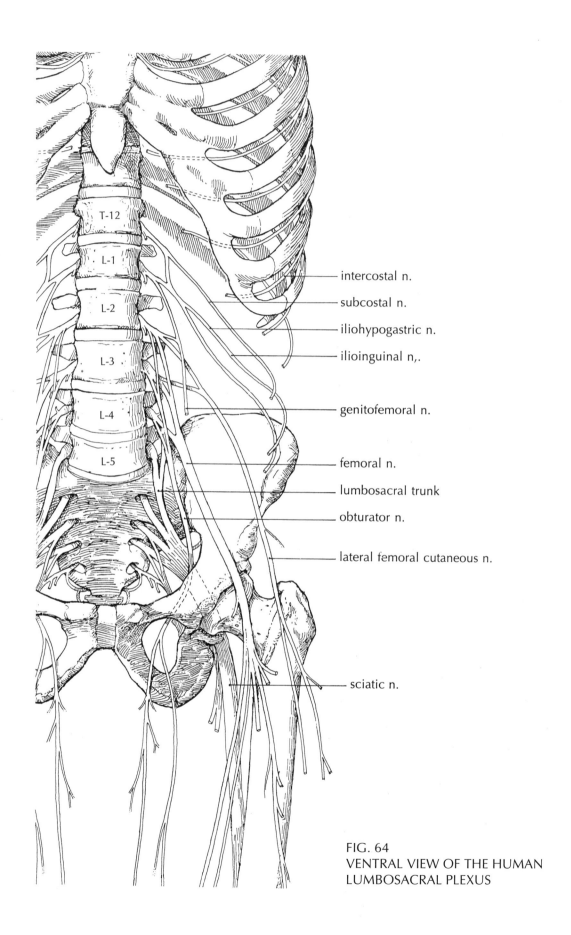

intercostal n.

subcostal n.

iliohypogastric n.

ilioinguinal n,.

genitofemoral n.

femoral n.

lumbosacral trunk

obturator n.

lateral femoral cutaneous n.

sciatic n.

T-12
L-1
L-2
L-3
L-4
L-5

FIG. 64
VENTRAL VIEW OF THE HUMAN
LUMBOSACRAL PLEXUS

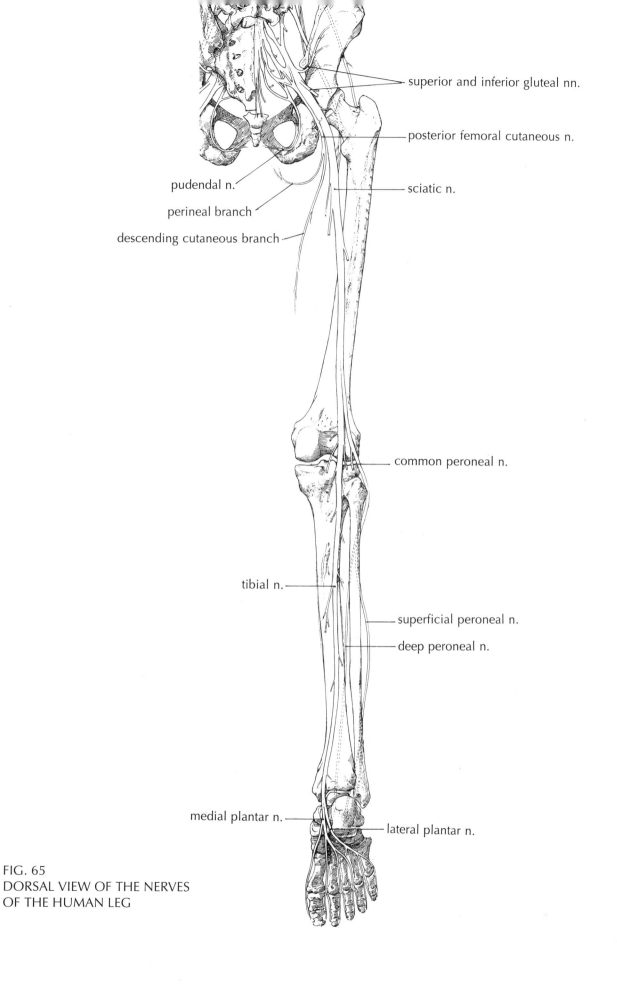

superior and inferior gluteal nn.

posterior femoral cutaneous n.

pudendal n.

sciatic n.

perineal branch

descending cutaneous branch

common peroneal n.

tibial n.

superficial peroneal n.

deep peroneal n.

medial plantar n.

lateral plantar n.

FIG. 65
DORSAL VIEW OF THE NERVES
OF THE HUMAN LEG

THE EYE

TUNICS OF THE EYE

Outer fibrous tunic

Cornea. The transparent anterior surface of the outer tunic.

Sclera. The opaque white membrane which covers the posterior part of the eyeball and serves to maintain the shape of the eye.

Middle vascular tunic

Choroid. The pigmented vascular layer which lies between the sclera and the retina.

Ciliary body. A structure composed of the thickened vascular tunic and the ciliary muscle. It gives attachment to the suspensory ligament of the lens.

Iris. A circular, contractile structure which is continuous with ciliary body. It lies between the cornea and the lens.

Tapetum lucidum. An iridescent layer within the choroid. Found in certain nocturnal mammals, it consists of organic crystals which serve to enhance night vision.

Internal tunic

Fovea centralis. The central depression in the macula.

Macula. The oval yellowish area of the retina in which vision is most acute.

Optic disk. The point at which the retina is continuous with the optic nerve. It is insensitive to light.

Ora serrata. The anterior limit of the photoreceptive part of the retina.

Retina. The thin nervous membrane upon which images are projected by light passing through the lens. Posteriorly it is continuous with the optic nerve.

CHAMBERS AND REFRACTING MEDIA

Anterior chamber. The space bounded anteriorly by the cornea and posteriorly by the iris and the lens.

Aqueous humor. The clear fluid contained in the anterior and posterior chambers.

Lens. The transparent crystalline structure which serves to focus an image on the retina.

Posterior chamber. The space bounded anteriorly by the iris and posteriorly by the suspensory ligament of the lens.

Suspensory ligament of the lens. Fibrils which connect the ciliary body to the lens.

Vitreous body. The transparent semi-gelatinous substance which fills the interior of eyeball posterior to the lens.

Vitreous chamber. The space which contains the vitreous body.

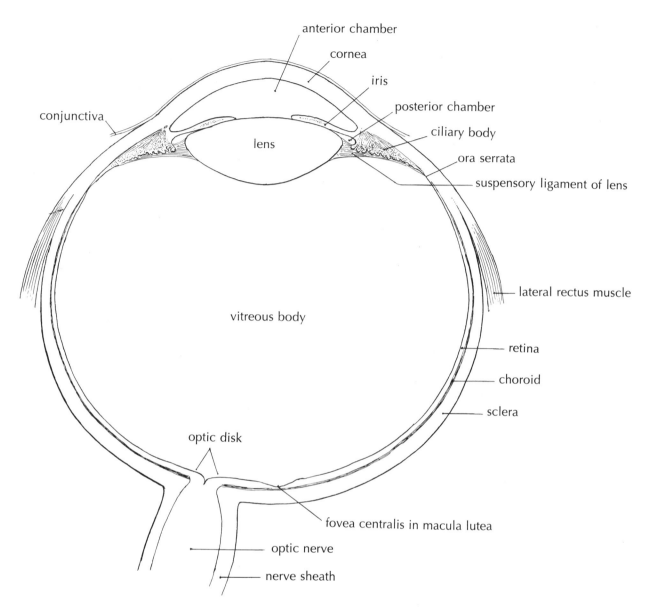

anterior chamber

cornea

iris

posterior chamber

ciliary body

ora serrata

suspensory ligament of lens

conjunctiva

lens

lateral rectus muscle

retina

choroid

sclera

vitreous body

optic disk

fovea centralis in macula lutea

optic nerve

nerve sheath

FIG. 66
SCHEMATIC HORIZONTAL SECTION OF
THE RIGHT HUMAN EYEBALL, AS SEEN
FROM ABOVE (after Spalteholz)

Muscle	Origin	Insertion	Nerve
Inferior oblique	maxillary bone	sclera near tendon of lateral rectus	oculomotor (III)
Lateral rectus	bone around optic foramen	sclera	abducens (VI)
Retractor oculi	bone around optic foramen	sclera	abducens (VI)
Superior oblique	cranial border of optic foramen	passes through fibrous band (trochlea); inserts on sclera near superior rectus	trochlear (IV)
Superior rectus **Medial rectus** **Inferior rectus**	bone around optic foramen	sclera	oculomotor (III)

ACCESSORY ORGANS OF THE EYE

Conjunctiva. The mucous membrane of the eye. It lines the nictitating membrane and the inner surfaces of the lids and is reflected onto the anterior part of the sclera.

Lacrimal canals and nasolacrimal duct. Passages which convey tears from the eye to the nasal cavity.

Lacrimal gland. A gland which lies on the surface of the eyeball beneath the lateral angle. It secretes tears.

Levator palpebrae superioris. A muscle which originates near the optic foramen and inserts on the upper eyelid. It is innervated by the oculomotor nerve and acts to raise the lid.

Nictitating membrane. The semitransparent membrane at the medial corner of the eye. Present in certain mammals but vestigial in humans, it serves to keep the eye clean and moist.

Orbicularis oculi. The sphincter muscle which closes the eyelids.

Periorbita. The membranous sac which encloses the eyeball together with its muscles and glands.

Trochlea. A ring-shaped fibrous band attached to the frontal bone. The tendon of the superior oblique passes through it.

Remove the zygomatic arch, the ramus of the dentary bone, and the muscles of the jaw as necessary to expose the contents of the orbit. Identify the structures illustrated in Figure 67.

On the right side, trim away the dorsal part of the orbit to expose the muscles as seen in Figure 68. On the left side, cut the ocular muscles near their insertions, cut the optic nerve, and remove the eyeball. Make a horizontal cut around the eyeball and remove the dorsal half to expose the structures illustrated in Figure 68.

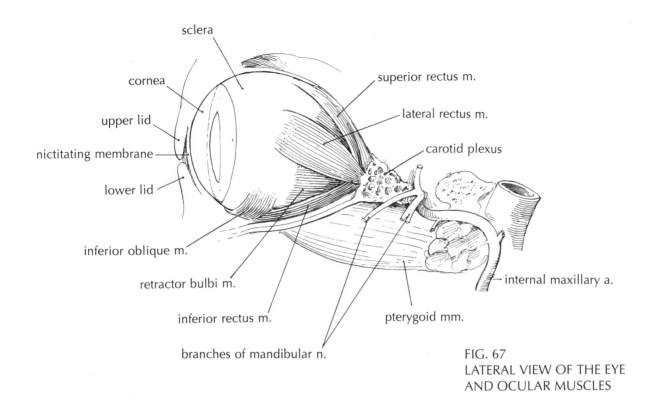

sclera

cornea

upper lid

nictitating membrane

lower lid

inferior oblique m.

retractor bulbi m.

inferior rectus m.

branches of mandibular n.

superior rectus m.

lateral rectus m.

carotid plexus

internal maxillary a.

pterygoid mm.

FIG. 67
LATERAL VIEW OF THE EYE
AND OCULAR MUSCLES

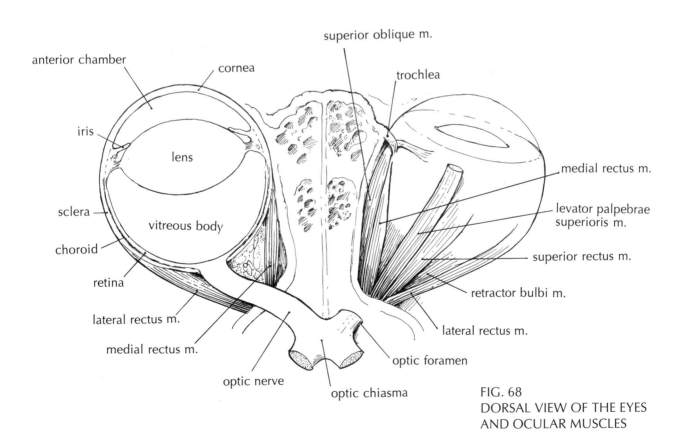

superior oblique m.

trochlea

anterior chamber

cornea

iris

lens

sclera

vitreous body

choroid

retina

lateral rectus m.

medial rectus m.

optic nerve

optic chiasma

optic foramen

lateral rectus m.

retractor bulbi m.

superior rectus m.

levator palpebrae
superioris m.

medial rectus m.

FIG. 68
DORSAL VIEW OF THE EYES
AND OCULAR MUSCLES

THE EAR

EXTERNAL EAR

External acoustic meatus. The cartilaginous tube which extends from the pinna to the tympanic membrane.

Pinna. The visible portion of the external ear. It consists of a thin *auricular cartilage* covered by skin and attached to the skull by muscles.

MIDDLE EAR

Auditory ossicles. Three minute bones which transmit vibrations from the tympanic membrane to the inner ear:

Malleus (hammer). The auditory ossicle which is connected to the tympanic membrane.

Incus (anvil). The auditory ossicle which lies between the malleus and the stapes.

Stapes (stirrup). The auditory ossicle which fits into the fenestra vestibuli (an opening in the petrous part of the temporal bone).

Auditory (Eustachian) tube. The cartilaginous passage via which the tympanic cavity communicates with the nasopharynx.

Tympanic cavity. A cavity within the petrous part of the temporal bone.

Tympanic membrane. The thin semitransparent membrane which separates the external acoustic meatus from the tympanic cavity.

INTERNAL EAR

Endolymph. The clear fluid contained within the membranous labyrinth.

Fenestra cochleae (round window). The opening between the osseous labyrinth and the tympanic cavity. In life it is closed by the *secondary tympanic membrane.*

Fenestra vestibuli (oval window). The opening between the osseous labyrinth and tympanic cavity. In life it is closed by the base of stapes.

Membranous labyrinth. The organ of hearing and balance, consisting of a spiral part, the *cochlear duct;* two sacs, the *utricle* and the *saccule,* and three *semicircular ducts.* The membranous labyrinth lies within the osseous labyrinth and contains sensory areas which receive filaments of the vestibulocochlear nerve (VIII).

Osseous labyrinth. A series of cavities within the petrous part of the temporal bone. In dry bone, the osseous labyrinth communicates with the tympanic cavity via two openings: the fenestra vestibuli and the fenestra cochlea.

Perilymph. The clear fluid which occupies the space between the osseous labyrinth and the membranous labyrinth.

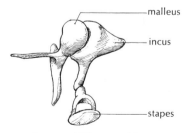

human middle ear ossicles
(after Sobotta)

Medial view of the right human middle ear ossicles (after Sobotta).

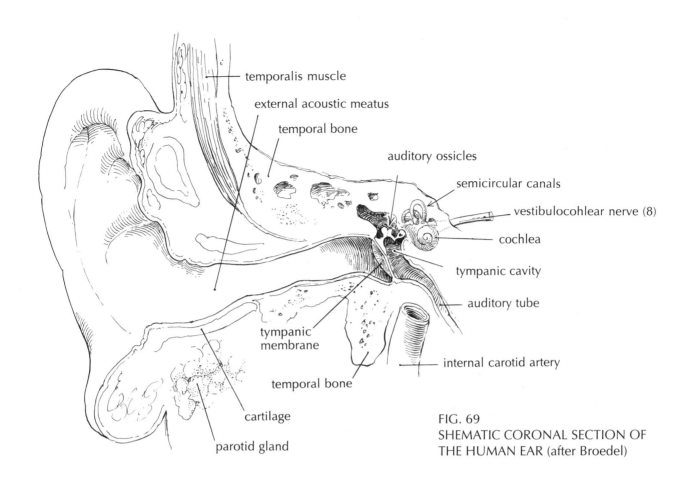

temporalis muscle

external acoustic meatus

temporal bone

auditory ossicles

semicircular canals

vestibulocohlear nerve (8)

cochlea

tympanic cavity

auditory tube

internal carotid artery

tympanic membrane

temporal bone

cartilage

parotid gland

FIG. 69
SHEMATIC CORONAL SECTION OF
THE HUMAN EAR (after Broedel)

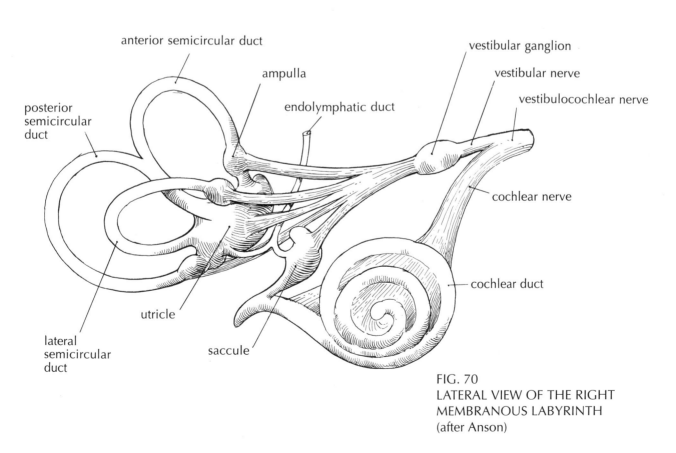

anterior semicircular duct

ampulla

endolymphatic duct

vestibular ganglion

vestibular nerve

vestibulocochlear nerve

posterior
semicircular
duct

cochlear nerve

lateral
semicircular
duct

utricle

saccule

cochlear duct

FIG. 70
LATERAL VIEW OF THE RIGHT
MEMBRANOUS LABYRINTH
(after Anson)

Definitions of Descriptive Terms

Anterior, **cephalic**, or **cranial**. In comparative anatomy: toward the head. In human anatomy: the term anterior is also used as synonym for ventral.

Caudal. Toward the tail.

Deep. Toward the middle of the trunk or of a limb.

Distal. Pertaining to a position removed from the center of the body or from the origin of a structure.

Dorsal. Toward the vertebral column or back.

Horizontal or frontal plane. A plane which lies at right angles to the median sagittal plane and parallel to the dorsal and ventral surfaces of the body.

Inferior. Below.

Lateral. Pertaining to the side of the body.

Medial. Pertaining to the middle, or midline of the body.

Median sagittal plane. The plane which divides the body into right and left halves.

Posterior. Toward the back or dorsal surface of the body.

Proximal. Pertaining to a position close to the center of the body or to the origin of a structure.

Right and **left** are determined with reference to the orientation of the specimen, not with reference to the orientation of the observer.

Sagittal plane. Any plane parallel to the median sagittal plane.

Superficial. Toward the surface of a trunk or of a limb.

Superior. Above.

Transverse or cross plane. The plane which lies at right angles to both the sagittal and the horizontal planes.

Ventral. Toward the abdomen or underside of the body.

Median sagittal plane

Horizontal or *frontal plane*

Transverse or *cross plane*

Bibliography

Anson, Barry, and James A. Donaldson. *Surgical Anatomy of the Temporal Bone and Ear.* Philadelphia: Saunders, 1973.

Broedel, Max. *Three Unpublished Drawings of the Anatomy of the Human Ear.* Philadelphia: Saunders, 1946.

Crouch, James E. *Text Atlas of Cat Anatomy.* Illustrated by Martha B. Lackey. Philadelphia: Lea and Febiger, 1969.

Dorland, W. A. *Dorland's Illustrated Medical Dictionary.* 27th ed. Philadelphia: W. B. Saunders, 1988.

Gray, Henry. *Gray's Anatomy.* 28th American ed. Edited by Charles Mayo Goss. Philadelphia: Lea and Febiger, 1966.

Gray, Henry. *Gray's Anatomy.* 36th English ed. Edited by Peter Williams and Roger Warwick. London: Churchill Livingstone, 1980.

Mivart, St. George. *The Cat: An Introduction to the Study of Backboned Animals, Especially Mammals.* London: J. Murray, 1881.

Reighard, Jacob, and H. S. Jennings. *Anatomy of the Cat.* 3rd ed. New York: Holt, Rinehart, and Winston, 1935.

Spalteholz, Werner. *Hand-Atlas of Human Anatomy.* Philadelphia: J. B. Lippincott, 1923.

Straus-Durckheim, Hercule. *Anatomie descriptive et comparative du chat.* Paris, 1845.

Walker, Warren F., and Dominique G. Homburger. *Vertebrate Dissection.* 8th ed. Philadelphia: Saunders College Publishing, 1992.

Index

abdominal cavity, 31
abductor, 10
adductor, 10
alveoli, 28
amphiarthrosis, 3
anterior chamber of eye, 80
aortic valve, 55
aortic vestibule, 55
aponeurosis, 10
aqueous humor, 80
arbor vitae, 65
ARTERIES
    abdominal aorta, 52
    axillary, 51
    brachial, 51
    brachiocephalid, 51
    celiac, 52
    common carotid, 51
    coronary, 51
    external iliac, 52
    hepatic, 52
    iliolumbar, 52
    inferior mesenteric, 52
    internal carotid, 51
    left gastric, 52
    lumbar, 52
    ovarian, 52
    phrenicoabdominal, 52
    renal, 52
    splenic, 52
    subclavian, 51
    superior mesenteric, 52
    testicular, 52
    thoracic aorta, 51
articular cartilage, 3
atria, 49
atrioventiruclar opening, 54, 55
attachment of muscle, 10
auditory ossicles, 84
auditory tube, 84
auricles, 49
autonomic nervous system, 62, 74

basal ganglia, 64
belly of muscle, 10
bicuspid valve, 55
bile duct, 32
bone marrow, 3
BONES
    basisphenoid, 5
    dentary, 4
    ethmoid, 4, 5
    femur, 9
    fibula, 9
    frontal, 4
    humerus, 8
    innominate bone, 9
    interparietal, 4
    maxilla, 4, 5
    metacarpals, 8
    metatarsals, 9
    nasal, 4
    occipital, 4, 5
    palatine, 4, 5
    parietal, 4
    premaxilla, 4, 5
    presphenoid, 5
    radius, 8
    scapula, 8
    tarsals, 9
    temporal, 4
    tibia, 9
    turbinate, 5
    ulna, 8
    vertebrae, 6; 7
    vomer, 5
broad ligament of uterus, 42
bronchi, 28
bulbourethral gland, 40

caecum, 32
cancellous bone, 3
carpals, 8
cartilage, 3
central canal of spinal cord, 62
central nervous system, 62

cerebellar peduncles, 65
cerebellum, 65
cerebral aqueduct, 65
cerebral cortex, 64
cerebral peduncles, 65
cerebrospinal fluid, 62
cerebrum, 64
cervix, 42
chambers of eye, 80
chondrocranium, 5
chordae tendineae, 54
choroid of eye, 80
ciliary body, 80
clitoris, 42
colon, 32
compact bone, 3
conjunctiva, 82
conus arteriosus, 49
conus medullaris, 62
cornea, 80
coronary sinus, 49
corpora cavernosa, 40
corpora quadrigemina, 65
corpus callosum, 64
corpus spongiosum, 40
costal cartilage, 3
cranial nerves, 66

depressor, defined, 10
diaphragm, 31
diarthrosis, 3
diencephalon, 64
dilator, defined, 10
dorsal ramus of spinal nerve, 63
dorsal root of spinal nerve, 63
ductus deferens, 40
duodenal papilla, 32
duodenum, 32
dura mater, 62

endolymph, 84
epididymis, 40

epiglottis, 30
epithalamus, 64
esophagus, 32
Eustachian tube, 84
extensor, 10
external auditory canal, 84
external inguinal ring, 40
external uterine orifice, 42
eye, 80

Fallopian tube, 42
fascia, defined, 10
femur, 9
fenestra cochleae, 84
fenestra vesibuli, 84
fibula, 9
filum terminale, 62
fimbriae, 42
flexor, defined, 10
fossa ovalis, 54
fovea, 80

gallbladder, 32
genioglossus m., 30
glans penis, 40
glottis, 30
gray matter, 62
greater omentum, 31

hard palate, 30
heart, 49
hilus of kidney, 38
humerus, 8
hypothalamus, 64

ileocolic valve, 32
ileum, 32
incus, 84
inguinal canal, 40
innominate bone, 9
insertion of muscle, defined, 10
interatrial septum,
internal inguinal ring, 40

internal tunic of eye, 80
interventricular septum, 55
intervertebral disk, 3
iris, 80

jejunum, 32
joints, types of, 3

kidney, 38

labia majora, 42
lacrimal canals, 82
lacrimal gland, 82
large intestine, 32'
larynx, 30
lateral ligament of bladder, 38
lens, 80
lesser omentum, 31
levator, defined, 10
ligament of ovary, 42
ligament, 3
ligamentum arteriosum, 49
liver, 32
lungs, 28

macula, 80
malleus, 84
median ligament of bladder, 38
mediastinum, 28
medulla oblongata, 65
medulla, 38
medullary cavity, 3
membrane bones of skull, 4
membranous labyrinth, 84
meninges, 62
mesencephalon, 65
mesenteries, 31
mesentery proper, 31
mesocolon, 31
mesoduodenum, 31
mesovarium, 42
metacarpals, 8
metatarsals, 9
metencephalon, 65
middle vascular tunic of eye, 80
mitral valve, 55
MUSCLES
    acromiodeltoid, 16
    acromiotrapezius, 16
    adductor femoris, 21
    adductor longus, 21
    adductor pollicis longus, 18

anconeus, 18
biceps brachii, 19
biceps femoris, 20
brachialis, 18
brachioradialis, 17
clavobrachialis, 16
clavotrapezius, 16
coracobrachialis, 19
epitrochlearis, 15
extensor carpi radialis longus & brevis, 17
extensor carpi ulnaris, 17
extensor digitorum communis, 17
extensor digitorum lateralis, 17
extensor digitorum longus, 20
extensor dorsi communis, 22
extensors of 1st & 2nd digits, 18
external oblique, 22
flexor carpi radialis, 15
flexor carpi ulnaris, 15
flexor digitorum longus, 21
flexor digitorum profundus, 19
flexor digitorum superficialis, 19
flexor hallucis longus, 21
gastrocnemius, 20
gracilis, 21
iliocostalis, 22
iliopsoas, 21
infraspinatus, 18
internal oblique, 22
latissimus dorsi, 17
levator scapulae ventralis, 16
longissimus dorsi, 22
masseter, 16
multifidus spinae, 22
oblique, of eye, 82
orbicularis oculi, 82
palmaris longus, 15
pectineus, 21
pectoralis major, 15
pectoralis minor, 15
peroneus brevis, 20
peroneus longus, 20
peroneus tertius, 20
plantaris, 21
pronator quadratus, 19
pronator teres, 15

quadriceps femoris, 21
rectractor oculi, 82
rectus abdominis, 22
rectus, of eye, 82
sartorius, 21
soleus, 20
spinalis dorsi, 22
spinodeltoid, 16
spinotrapezius, 17
sternomastoid, 16
subscapularis, 19
supinator, 18
supraspinatus, 18
temporal, 16
tensor fasciae latae, 20
teres major, 19
teres minor, 18
tibialis anterior, 20
tibialis posterior, 21
transversus, 22
triceps brachii , 17
vastus medialis, 21
xiphihumeralis, 15
musculi pectinati, 54
myelencephalon, 65

nasal cavity, 30
nasolacrimal duct, 82
neopallium, 64
nephron, 38
NERVES
    abducens, 66
    axillary, 70
    cervical, 70
    cochlear, 66
    facial, 66
    femoral, 72
    genitofemoral, 72
    glossopharyngeal, 66
    hypoglossal, 66
    inferior gluteal, 72
    inferior hemorrhoidal, 72
    lateral femoral cutaneous, 72
    long thoracic, 70
    mandibular, 66
    maxillary, 66
    median, 70
    musculocutaneous, 70
    obturator, 72
    oculomotor, 66
    olfactory, 66
    ophthalmic, 66
    optic, 66
    pectoral, 70
    phrenic, 70

posterior femoral cutaneous, 72
    pudendal, 72
    radial, 70
    sciatic, 72
    spinal accessory, 66
    subscapular, 70
    superior gluteal, 72
    suprascapular, 70
    thoracic, 70
    trigeminal, 66, 68
    trochlear, 66
    ulnar, 70
    vagus, 66, 69
    vestibular, 66
    vestibulocochlear, 66
nictitating membrane, 82
nutrient foramen, 3

olfactory mucosa, 30
omental bursa, 31
opening of auditory tube, 30
opening of coronary sinus, 54
optic disk, 80
ora serrata, 80
oral cavity, 30
origin of a muscle, defined, 10
osseous labyrinth, 84
outer fibrious tunic of eye, 80
oval window, 84
ovary, 42

palatine tonsils, 30
pancreas, 32
pancreatic ducts, 32
parasympathetic system, 74
pelvis, 9
pericardium, 49
perilymph, 84
periorbita, 82
periosteum, 3
peripheral nervous system, 62
peritoneal cavity, 31
peritoneum, 31
Peyer's patches, 32
pharynx, 30
pia mater, 62
pinna, 84
pleura, 28
pleural cavity, 28
pons, 65

posterior chamber of eye, 80
prepuce, 40
pronator, defined, 10
prostate gland, 40
pulmonary arteries, 49
pulmonary ligament, 28
pulmonary trunk, 49
pulmonary valve, 54
pyloric valve, 32
pylorus, 32

radius, 8
rami communicantes, 74
rectovesical pouch, 38
rectum, 32
renal capsule, 38
renal cortex, 38
renal papilla, 38
renal pelvis, 38
renal sinus, 38
retina, 80
retroperitoneal, 31
rhinencaphalon, 64
round ligament of uterus, 42
round window, 84

scapula, 8
sclera, 80
scrotum, 40
skull, 4, 5
small intestine, 32
soft palate, 30
spermatic cord, 40
sphincter, defined, 10
spinal cord, 62
spinal nerve, 63
spleen, 32
stapes, 84
stomach, 32
suprarenal gland, 38
suspensory ligament of lens, 80
sympathetic system, 74
sympathetic trunks, 74
synarthrosis, 3
synovial fluid, 3
synovial joint, 3
synovial membrane, 3

tapetum lucidum, 80
tarsals, 9
telencephalon, 64
tendon, defined, 10

testis, 40
thalamus, 64
tibia, 9
trabeculae carneae, 54
trachea, 28
tricuspid valve, 54
tunics of eye, 80
tympanic cavity, 84
tympanic membrane, 84

ulna, 8
ureter, 38
urethra, 38
urinary bladder, 38
urogenital aperture, 42
urogenital sinus, 42
uterine tube, 42
uterus, 42
vagina, 42
vasa efferentia testis, 40
VEINS
    azygos, 50
    brachiocephalic, 50
    external jugular, 50
    hepatic, 52
    internal jugular, 50
    internal thoracic, 50

portal, 52
postcava, 52
precava, 50
subclavian, 50
ventral ramus of spinal nerve, 63
ventral root of spinal nerve, 63
ventricles, 49
vertebrae, 6, 7
vesicouterine pouch, 38
vitreous body, 80
vocal cords, 30

white matter, 62, 64